SCIENTIFIC AMERICAN

OIL AND THE FUTURE OF ENERGY:

Climate Repair • Hydrogen • Nuclear Fuel
Renewable and Green Sources • Energy Efficiency

The Editors of *Scientific American*

THE LYONS PRESS

Guilford, Connecticut
An imprint of The Globe Pequot Press

CONTENTS

PART II: THE FUTURE OF ENERGY

OIL AND THE
FUTURE OF ENERGY

PART ONE

THE ERA OF
(OIL AND COAL)

(1)

OIL PRODUCTION, CLIMATE CHANGE AND GEOPOLITICS

The End of Cheap Oil

Global production of conventional oil will begin to decline sooner than most people think, probably within 10 years.

COLIN J. CAMPBELL AND JEAN H. LAHERRÈRE

In 1973 and 1979 a pair of sudden price increases rudely awakened the industrial world to its dependence on cheap crude oil. Prices first tripled in response to an Arab embargo and then nearly doubled again when Iran dethroned its shah, sending the major economies sputtering into recession. Many analysts warned that these crises proved that the world would soon run out of oil. Yet they were wrong.

Their dire predictions were emotional and political reactions; even at the time, oil experts knew that they had no scientific basis. Just a few years earlier oil explorers had discovered enormous new oil provinces on the North Slope of Alaska and below the North Sea off the coast of Europe. By 1973 the world had consumed, according to many experts' best estimates, only about one-eighth of its endowment of readily accessible crude

oil (so-called conventional oil). The five Middle Eastern members of the Organization of Petroleum Exporting Countries (OPEC) were able to hike prices not because oil was growing scarce but because they had managed to corner 36 percent of the market. Later, when demand sagged, and the flow of fresh Alaskan and North Sea oil weakened OPEC's economic stranglehold, prices collapsed.

The next oil crunch will not be so temporary. Our analysis of the discovery and production of oil fields around the world suggests that within the next decade, the supply of conventional oil will be unable to keep up with demand. This conclusion contradicts the picture one gets from oil industry reports, which boasted of 1,020 billion barrels of oil (Gbo) in "proved" reserves at the start of 1998. Dividing that figure by the current production rate of about 23.6 Gbo a year might suggest that crude oil could remain plentiful and cheap for 43 more years—probably longer, because official charts show reserves growing.

Unfortunately, this appraisal makes three critical errors. First, it relies on distorted estimates of reserves. A second mistake is pretending that production will remain constant. Third and most important, conventional wisdom erroneously assumes that the last bucket of oil can be pumped from the ground just as quickly as the barrels of oil gushing from wells today. In fact, the rate at which any well—or any country—can produce oil always rises to a maximum and then, when about half the oil is gone, begins falling gradually back to zero.

From an economic perspective, when the world runs completely out of oil is thus not directly relevant: what matters is when production begins to taper off. Beyond that point, prices will rise unless demand declines commensurately. Global production of conventional oil will begin to decline sooner than most people think, probably within 10 years. Using several different techniques to estimate the current reserves of conventional oil and the amount still left to be discovered, we conclude that the decline will begin before 2010.

Digging for the True Numbers

We have spent most of our careers exploring for oil, studying reserve figures and estimating the amount of oil left to discover, first while employed at major oil companies and later as independent consultants. Over the years, we have come to appreciate that the relevant statistics are far more complicated than they first appear.

Consider, for example, three vital numbers needed to project future oil production. The first is the tally of how much oil has been extracted to date, a figure known as cumulative production. The second is an estimate of reserves, the amount that companies can pump out of known oil fields before

having to abandon them. Finally, one must have an educated guess at the quantity of conventional oil that remains to be discovered and exploited. Together they add up to ultimate recovery, the total number of barrels that will have been extracted when production ceases many decades from now.

The obvious way to gather these numbers is to look them up in any of several publications. That approach works well enough for cumulative production statistics because companies meter the oil as it flows from their wells. The record of production is not perfect (for example, the two billion barrels of Kuwaiti oil wastefully burned by Iraq in 1991 is usually not included in official statistics), but errors are relatively easy to spot and rectify. Most experts agree that the industry had removed just over 800 Gbo from the earth at the end of 1997.

Getting good estimates of reserves is much harder, however. Almost all the publicly available statistics are taken from surveys conducted by the *Oil and Gas Journal* and *World Oil*. Each year these two trade journals query oil firms and governments around the world. They then publish whatever production and reserve numbers they receive but are not able to verify them.

The results, which are often accepted uncritically, contain systematic errors. For one, many of the reported figures are unrealistic. Estimating reserves is an inexact science to begin with, so petroleum engineers assign a probability to their assessments. For example, if, as geologists estimate, there is a 90 percent chance that the Oseberg field in Norway contains 700 million barrels of recoverable oil but only a 10 percent chance that it will yield 2,500 million more barrels, then the lower figure should be cited as the so-called P90 estimate (P90 for "probability 90 percent") and the higher as the P10 reserves.

In practice, companies and countries are often deliberately vague about the likelihood of the reserves they report, preferring instead to publicize whichever figure, within a P10 to P90 range, best suits them. Exaggerated estimates can, for instance, raise the price of an oil company's stock.

The members of OPEC have faced an even greater temptation to inflate their reports because the higher their reserves, the more oil they are allowed to export. National companies, which have exclusive oil rights in the main OPEC countries, need not (and do not) release detailed statistics on each field that could be used to verify the country's total reserves. There is thus good reason to suspect that when, during the late 1980s, six of the 11 OPEC nations increased their reserve figures by colossal amounts, ranging from 42 to 197 percent, they did so only to boost their export quotas.

Previous OPEC estimates, inherited from private companies before governments took them over, had probably been conservative, P90 numbers. So some upward revision was warranted. But no major new discoveries or technological breakthroughs justified the addition of a staggering 287 Gbo. That increase is more than all the oil ever discovered in the U.S.—plus 40 percent. Non-OPEC countries, of course, are not above fudging their numbers either: 59 nations stated in 1997 that their reserves were unchanged from 1996. Because reserves naturally drop as old fields are drained and jump when new fields are discovered, perfectly stable numbers year after year are implausible.

Unproved Reserves

Another source of systematic error in the commonly accepted statistics is that the definition of reserves varies widely from region to region. In the U.S., the Securities and Exchange Commission allows companies to call reserves "proved" only if the oil lies near a producing well and there is "reasonable certainty" that it can be recovered profitably at current oil prices, using existing technology. So a proved reserve estimate in the U.S. is roughly equal to a P90 estimate.

Regulators in most other countries do not enforce particular oil-reserve definitions. For many years, the former Soviet countries have routinely released wildly optimistic figures—essentially P10 reserves. Yet analysts have often misinterpreted these as estimates of "proved" reserves. *World Oil* reckoned reserves in the former Soviet Union amounted to 190 Gbo in 1996, whereas the *Oil and Gas Journal* put the number at 57 Gbo. This large discrepancy shows just how elastic these numbers can be.

Using only P90 estimates is not the answer, because adding what is 90 percent likely for each field, as is done in the U.S., does not in fact yield what is 90 percent likely for a country or the entire planet. On the contrary, summing many P90 reserve estimates always understates the amount of proved oil in a region. The only correct way to total up reserve numbers is to add the mean, or average, estimates of oil in each field. In practice, the median estimate, often called "proved and probable," or P50 reserves, is more widely used and is good enough. The P50 value is the number of barrels of oil that are as likely as not to come out of a well during its lifetime, assuming prices remain within a limited range. Errors in P50 estimates tend to cancel one another out.

We were able to work around many of the problems plaguing estimates of conventional reserves by using a large body of statistics maintained by Petroconsultants in Geneva. This information, assembled over 40 years from myriad sources, covers some 18,000 oil fields worldwide. It, too, contains some dubious reports, but we did our best to correct these sporadic errors.

According to our calculations, the world had at the end of 1996 approximately 850 Gbo of conventional oil in P50 reserves—substantially less than the 1,019 Gbo reported in the *Oil and Gas Journal* and the 1,160 Gbo estimated by *World Oil.* The difference is actually greater than it appears because our value represents the amount most likely to come out of known oil fields, whereas the larger number is supposedly a cautious estimate of proved reserves.

For the purposes of calculating when oil production will crest, even more critical than the size of the world's reserves is the size of ultimate recovery—all the cheap oil there is to be had. In order to estimate that, we need to know whether, and how fast, reserves are moving up or down. It is here that the official statistics become dangerously misleading.

Diminishing Returns

According to most accounts, world oil reserves have marched steadily upward over the past 20 years. Extending that apparent trend into the future, one could easily conclude, as the U.S. Energy Information Administration has, that oil production will continue to rise unhindered for decades to come, increasing almost two thirds by 2020.

Such growth is an illusion. About 80 percent of the oil produced today flows from fields that were found before 1973, and the great majority of them are declining. In the 1990s oil companies have discovered an average of seven Gbo a year; last year they drained more than three times as much. Yet official figures indicated that proved reserves did not fall by 16 Gbo, as one would expect—rather they *expanded* by 11 Gbo. One reason is that several dozen governments opted not to report declines in their reserves, perhaps to enhance their political cachet and their ability to obtain loans. A more important cause of the expansion lies in revisions: oil companies replaced earlier estimates of the reserves left in many fields with higher numbers. For most purposes, such amendments are harmless, but they seriously distort forecasts extrapolated from published reports.

To judge accurately how much oil explorers will uncover in the future, one has to backdate every revision to the year in which the field was first discovered—not to the year in which a company or country corrected an earlier estimate. Doing so reveals that global discovery peaked in the early 1960s and has been falling steadily ever since. By extending the trend to zero, we can make a good guess at how much oil the industry will ultimately find.

We have used other methods to estimate the ultimate recovery of conventional oil for each country [see sidebar, "How Much Oil Is Left to Find?" page 7], and we calculate that the oil industry will be able to recover only about another 1,000 billion barrels of conventional oil. This number, though great, is little more than the 800 billion barrels that have already been extracted.

It is important to realize that spending more money on oil exploration will not change this situation. After the price of crude hit all-time highs in the early 1980s, explorers developed new technology for finding and recovering oil, and they scoured the world for new fields. They found few: the discovery rate continued its decline uninterrupted. There is only so much crude oil in the world, and the industry has found about 90 percent of it.

Predicting the Inevitable

Predicting when oil production will stop rising is relatively straightforward once one has a good estimate of how much oil there is left to produce. We simply apply a refinement of a technique first published in 1956 by M. King Hubbert. Hubbert observed that in any large region unrestrained extraction of a finite resource rises along a bell-shaped curve that peaks when about half the resource is gone. To demonstrate his theory, Hubbert fitted a bell curve to production statistics and projected that crude oil production in the lower 48 U.S. states would rise for 13 more years, then crest in 1969, give or take a year. He was right: production peaked in 1970 and has continued to follow Hubbert curves with only minor deviations. The flow of oil from several other regions, such as the former Soviet Union and the collection of all oil producers outside the Middle East, also follows Hubbert curves quite faithfully.

The global picture is more complicated, because the Middle East members of OPEC deliberately reined back their oil exports in the 1970s, while other nations continued producing at full capacity. Our analysis reveals that a number of the largest producers, including Norway and the UK, will reach their peaks around the turn of the millennium unless they sharply curtail production. By 2002 or so the world will rely on Middle East nations, particularly five near the Persian Gulf (Iran, Iraq, Kuwait, Saudi Arabia and the United Arab Emirates), to fill in the gap between dwindling supply and growing demand. But once approximately 900 Gbo have been consumed, production must soon begin to fall. Barring a global recession, it seems most likely that world production of conventional oil will peak during the first decade of the 21st century.

Perhaps surprisingly, that prediction does not shift much even if our estimates are a few hundred billion barrels high or low. Craig Bond Hatfield of the University of Toledo, for example, has conducted his own analysis based on a 1991 estimate by the U.S. Geological Survey of 1,550 Gbo remaining—55 percent higher than our figure. Yet he similarly concludes that the world will hit maximum oil production within the next 15 years. John D. Edwards of the University of Colorado published last August one of the most optimistic recent estimates of oil remaining: 2,036 Gbo. (Edwards concedes that the industry has only a 5 percent chance of attaining that very high goal.) Even so, his calculations suggest that conventional oil will top out in 2020.

Smoothing the Peak

Factors other than major economic changes could speed or delay the point at which oil production begins to decline. Three in particular have often led economists and academic geologists to dismiss concerns about future oil production with naive optimism.

First, some argue, huge deposits of oil may lie undetected in far-off corners of the globe. In fact, that is very unlikely. Exploration has pushed the frontiers back so far that only extremely deep water and polar regions remain to be fully tested, and even their prospects are now reasonably well understood. Theoretical advances in geochemistry and geophysics have made it possible to map productive and prospective fields with impressive accuracy. As a result, large tracts can be condemned as barren. Much of the deepwater realm, for example, has been shown to be absolutely nonprospective for geologic reasons.

What about the much touted Caspian Sea deposits? Our models project that oil production from that region will grow until around 2010. We agree with analysts at the USGS World Oil Assessment program and elsewhere who rank the total resources there as roughly equivalent to those of the North Sea—that is, perhaps 50 Gbo but certainly not several hundreds of billions as sometimes reported in the media.

A second common rejoinder is that new technologies have steadily increased the fraction of oil that can be recovered from fields in a basin—the so-called recovery factor. In the 1960s oil companies assumed as a rule of thumb that only 30 percent of the oil in a field was typically recoverable; now they bank on an average of 40 or 50 percent. That progress will continue and will extend global reserves for many years to come, the argument runs.

Of course, advanced technologies will buy a bit more time before production starts to fall [see "Oil Production in the 21st Century," page 10]. But most of the apparent improvement in recovery factors is an artifact of reporting. As oil fields grow old, their owners often deploy newer technology to slow their decline. The falloff also allows engineers to gauge the size of the field more accurately and to correct previous underestimation—in particular

P90 estimates that by definition were 90 percent likely to be exceeded.

Another reason not to pin too much hope on better recovery is that oil companies routinely count on technological progress when they compute their reserve estimates. In truth, advanced technologies can offer little help in draining the largest basins of oil, those onshore in the Middle East where the oil needs no assistance to gush from the ground.

Last, economists like to point out that the world contains enormous caches of unconventional oil that can substitute for crude oil as soon as the price rises high enough to make them profitable. There is no question that the resources are ample: the Orinoco oil belt in Venezuela has been assessed to contain a staggering 1.2 trillion barrels of the sludge known as heavy oil. Tar sands and shale deposits in Canada and the former Soviet Union may contain the equivalent of more than 300 billion barrels of oil. Theoretically, these unconventional oil reserves could quench the world's thirst for liquid fuels as conventional oil passes its prime. But the industry will be hard-pressed for the time and money needed to ramp up production of unconventional oil quickly enough.

Such substitutes for crude oil might also exact a high environmental price. Tar sands typically emerge from strip mines. Extracting oil from these sands and shales creates air pollution. The Orinoco sludge contains heavy metals and sulfur that must be removed. So governments may restrict these industries from growing as fast as they could. In view of these potential obstacles, our skeptical estimate is that only 700 Gbo will be produced from unconventional reserves over the next 60 years.

On the Down Side

Meanwhile, global demand for oil is currently rising at more than 2 percent a year. Since 1985, energy use is up about 30 percent in Latin America, 40 percent in Africa and 50 percent in Asia. The Energy Information Administration forecasts that worldwide demand for oil will increase 60 percent (to about 40 Gbo a year) by 2020.

The switch from growth to decline in oil production will thus almost certainly create economic and political tension. Unless alternatives to crude oil quickly prove themselves, the market share of the OPEC states in the Middle East will rise rapidly. Within two years, these nations' share of the global oil business will pass 30 percent, nearing the level reached during the oil-price shocks of the 1970s. By 2010 their share will quite probably hit 50 percent.

The world could thus see radical increases in oil prices. That alone might be sufficient to curb demand, flattening production for perhaps 10 years. (Demand fell more than 10 percent after the 1979 shock and took 17 years to recover.) But by 2010 or so, many Middle Eastern nations will themselves be past the midpoint. World production will then have to fall.

With sufficient preparation, however, the transition to the post-oil economy need not be traumatic. If advanced methods of producing liquid fuels from natural gas can be made profitable and scaled up quickly, gas could become the next source of transportation fuel [see "Liquid Fuels from Natural Gas," page 147]. Safer nuclear power, cheaper renewable energy and oil conservation programs could all help postpone the inevitable decline of conventional oil.

Countries should begin planning and investing now. In November a panel of energy experts appointed by President Bill Clinton strongly urged the administration to increase funding for energy research by $1 billion over the next five years. That is a small step in the right direction, one that must be followed by giant leaps from the private sector.

The world is not running out of oil—at least not yet. What our society does face, and soon, is the end of the abundant and cheap oil on which all industrial nations depend.

—MARCH 1998

How Much Oil Is Left to Find?

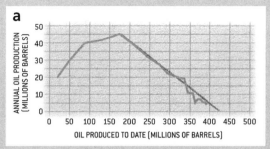

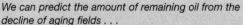

We can predict the amount of remaining oil from the decline of aging fields . . .

Wе combined several techniques to conclude that about 1,000 billion barrels of conventional oil remain to be produced. First, we extrapolated published production figures for older oil fields that have begun to decline. The Thistle field off the coast of Britain, for example, will yield about 420 million barrels (a). Second, we plotted the amount of oil discovered so far in some regions against the cumulative number of exploratory wells drilled there. Because larger fields tend to be found first—they are simply too large to miss—the curve rises rapidly and then flattens, eventually reaching a theoretical maximum: for Africa, 192 Gbo. But the time and cost of exploration impose a more practical limit of perhaps 165 Gbo (b). Third, we analyzed the distribution of oil-field sizes in the Gulf of Mexico and other provinces. Ranked according to size and then graphed on a logarithmic scale, the fields tend to fall along a parabola that grows predictably over time (c). (Interestingly, galaxies, urban populations and other natural agglomerations also seem to fall along such parabolas.) Finally, we checked our estimates by matching our projections for oil production in large areas, such as the world outside the Persian Gulf region, to the rise and fall of oil discovery in those places decades earlier (d).

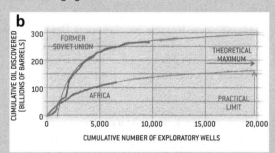

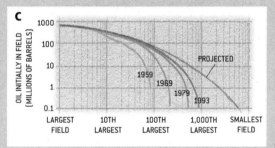

. . . from the diminishing returns on exploration in larger regions . . .

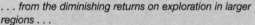

. . . by extrapolating the size of new fields into the future . . .

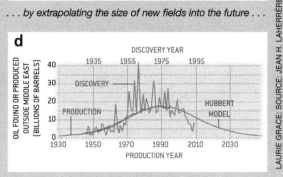

. . . and by matching production to earlier discovery trends.

LAURIE GRACE; SOURCE: JEAN H. LAHERRÈRE

The End of Oil

Will gas lines in the coming decade make those of 1973 look short?

PAUL RAEBURN

Review of *Hubbert's Peak: The Impending World Oil Shortage,*
by Kenneth S. Deffeyes, Princeton University Press, 2001

You have to wonder about the judgment of a man who writes, "As I drive by those smelly refineries on the New Jersey Turnpike, I want to roll the windows down and inhale deeply." But for Kenneth S. Deffeyes, that's the smell of home. The son of a petroleum engineer, he was born in Oklahoma, "grew up in the oil patch," became a geologist and worked for Shell Oil before becoming a professor at Princeton University. And he still knows how to wield a 36-inch-long pipe wrench.

In *Hubbert's Peak,* Deffeyes writes with good humor about the oil business, but he delivers a sobering message: the 100-year petroleum era is nearly over. Global oil production will peak sometime between 2004 and 2008, and the world's production of crude oil "will fall, never to rise again." If Deffeyes is right—and if nothing is done to reduce the increasing global thirst for oil—energy prices will soar, and economies will be plunged into recession as they desperately search for alternatives.

It's tempting to dismiss Deffeyes as just another of the doomsayers who have been predicting, almost since oil was discovered, that we are running out of it. But Deffeyes makes a persuasive

case that this time it's for real. This is an oilman and geologist's assessment of the future, grounded in cold mathematics. And it's frightening.

Deffeyes's prediction is based on the work of M. King Hubbert, a Shell geologist who in 1956 predicted that U.S. oil production would peak in the early 1970s and then begin to decline. Hubbert was dismissed by many experts inside and outside the oil industry. Pro-Hubbert and anti-Hubbert factions arose and persisted until 1970, when U.S. oil production peaked and started its long decline.

The Hubbert method is based on the observation that oil production in any region follows a bell-shaped curve. Production increases rapidly at first, as the cheapest and most readily accessible oil is recovered. As the difficulty of extracting the oil increases, it becomes more expensive and less competitive with other fuels. Production slows, levels off and begins to fall.

Hubbert demonstrated that total U.S. oil production in 1956 was tracing the upside of such a curve. To know when the curve would most likely peak, however, he had to know how much oil remained in the ground. Underground reserves provide a glimpse of the future: when the rate of new discoveries does not keep up with the growth of oil production, the amount of oil remaining underground begins to fall. That's a tip-off that a decline in production lies ahead.

Deffeyes used a slightly more sophisticated version of the Hubbert method to make the global calculations. The numbers pointed to 2003 as the year of peak production, but because estimates of global reserves are inexact, Deffeyes settled on a range from 2004 to 2008.

Three things could upset Deffeyes's prediction. One would be the discovery of huge new oil deposits. A second would be the development of drilling technology that could squeeze more oil from known reserves. And a third would be a steep rise in oil prices, which would make it profitable to recover even the most stubbornly buried oil.

In a delightfully readable and informative primer on oil exploration and drilling, Deffeyes addresses each point. First, the discovery of new oil reserves is unlikely—petroleum geologists have been nearly everywhere, and no substantial finds have been made since the 1970s. Second, billions have already been poured into drilling technology, and it's not going to get much better. And last,

even very high oil prices won't spur enough new production to delay the inevitable peak.

"This much is certain," he writes. "No initiative put in place starting today can have a substantial effect on the peak production year. No Caspian Sea exploration, no drilling in the South China Sea, no SUV replacements, no renewable energy projects can be brought on at a sufficient rate to avoid a bidding war for the remaining oil."

The only answer, Deffeyes says, is to move as quickly as possible to alternative fuels—including natural gas and nuclear power, as well as solar, wind and geothermal energy. "Running out of energy in the long run is not the problem," Deffeyes explains. "The bind comes during the next 10 years: getting over our dependence on crude oil."

The petroleum era is coming to a close. "Fossil fuels are a one-time gift that lifted us up from subsistence agriculture and eventually should lead us to a future based on renewable resources," Deffeyes writes. Those are strong words for a man raised in the oil patch. For the rest of us, the end of the world's dependence on oil means we need to make some tough political and economic choices. For Deffeyes, it means he can't go home again.

—OCTOBER 2001

Oil Production in the 21st Century

Recent innovations in underground imaging, steerable drilling and deepwater oil production could recover more of what lies below.

ROGER N. ANDERSON

On the face of it, the outlook for conventional oil—the cheap, easily recovered crude that has furnished more than 95 percent of all oil to date—seems grim. In 2010, according to forecasts, the world's oil-thirsty economies will demand about 10 billion more barrels than the industry will be able to produce. A supply shortfall that large, equal to almost half of all the oil extracted in 1997, could lead to price shocks, economic recession and even wars.

Fortunately, four major technological advances are ready to fill much of the gap by accelerating the discovery of new oil reservoirs and by dramatically increasing the fraction of oil within existing fields that can be removed economically, a ratio known as the recovery factor. These technologies could lift global oil production rates more than 20 percent by 2010 if they are deployed as planned on the largest oil fields within three to five years. Such rapid adoption may seem ambitious for an industry that traditionally has taken 10 to 20 years to put new inventions to use. But in this case, change will be spurred by formidable economic forces.

For example, in the past two years, the French oil company Elf has discovered giant deposits off the coast of West Africa. In the same period, the

company's stock doubled, as industry analysts forecasted that Elf's production would increase by 8 percent in 2001. If the other major oil producers follow suit, they should be able by 2010 to provide an extra five billion barrels of oil each year, closing perhaps half the gap between global supply and demand.

This article will cover the four advances in turn, beginning with a new way to track subterranean oil.

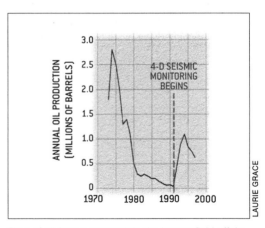

Flow of oil from a reservoir in the largest field off the Louisiana shore resurged in 1992, shortly after operators began using 4-D seismic monitoring to locate hidden caches of oil.

Tracking Oil in Four Dimensions

Finding oil became much more efficient after 1927, when geologists first successfully translated acoustic reflections into detailed cross sections of the earth's crust. Seismologists later learned how to piece together several such snapshots to create three-dimensional models of the oil locked inside layers of porous rock. Although this technique, known as 3-D seismic analysis, took more than a decade to become standard practice, it is now credited with increasing oil discovery and recovery rates by 20 percent.

In recent years, scientists in my laboratory at Columbia University and elsewhere have developed even more powerful techniques capable of tracking the movement of oil, gas and water as drilled wells drain the subterranean strata—a "4-D" scheme that includes the added dimension of time. This information can then be used to do a "what if" analysis on the oil field, designing ways to extract as much of the oil as quickly and cheaply as possible.

Compared with its predecessor, the 4-D approach seems to be catching on quickly: the number of oil fields benefiting from it has doubled in each of the past four years and now stands at about 60. Such monitoring can boost recovery factors by 10 to 15 percentage points. Unfortunately, the technique will work in only about half the world's major fields, those where relatively soft rock is suffused with oil and natural gas.

SEE *Figure 1 in color section.*

Gassing Things Up

When geologists began studying the new time-lapse measurements, they were surprised to discover that one of the most basic notions about oil movement—that it naturally settles between lighter gas above and heavier groundwater below—oversimplifies the behavior of real oil fields. In fact, most wells produce complex, fractal drainage patterns that cause the oil to mix with gas and water. As a result, specialists now know that the traditional technique of pumping a well until the oil slows to a trickle often leaves 60 percent or more of the oil behind.

A more efficient strategy is to pump natural gas, steam or liquid carbon dioxide into dead wells. The infusion then spreads downward through pores in the rock and, if one has planned carefully, pushes oil that otherwise would have been abandoned toward a neighboring well. Alternatively, water is often pumped below the oil to increase its pressure, helping it flow up to the surface.

Injections of steam and carbon dioxide have been shown to increase recovery factors by 10 to 15 percentage points. Unfortunately, they also raise the cost of oil production by 50 to 100 percent—and that added expense falls on top of a 10 to 25 percent surcharge for 4-D seismic monitoring. So unless carbon dioxide becomes much cheaper (perhaps because global-warming treaties restrict its release), these techniques will probably continue to serve only as a last resort.

Steering to Missed Oil

A third major technological advance, known as directional drilling, can tap bypassed deposits of oil at less expense than injection. Petroleum engineers can use a variety of new equipment to swing a well from vertical to entirely horizontal within a reservoir several kilometers underground.

Traditionally, drillers rotated the long steel pipe, or "string," that connects the rig at the surface to the bit at the bottom of the well. That method fails when the pipe must turn a corner—the bend would break the rotating string. So steerable drill strings do not rotate; instead, a mud-driven motor inserted near the bit turns only the diamond-tipped teeth that do the digging. An elbow of pipe placed between the mud motor and the bit controls the direction of drilling.

Threading a hole through kilometers of rock into a typical oil zone 30 meters (about 100 feet) thick is precise work. Schlumberger, Halliburton and other international companies have developed sophisticated sensors that significantly improve the accuracy of drilling. These devices, which operate at depths of up to 6,000 meters and at temperatures as high as 200 degrees Celsius (400 degrees Fahrenheit), attach to the drill pipe just above or below the mud motor. Some measure the electrical resistance of the surrounding rock. Others send out neutrons and gamma rays; then they count the number that are scattered back by the rock and pore fluids. These measurements and the current position of the bit (calculated by an inertial guidance system) are sent back to the surface through pulses in the flow of the very mud used to turn the motor and lubricate the well bore. Engineers can adjust the path of the drill accordingly, thus snaking their way to the most oil-rich part of the formation.

Once the hole is completed, drillers typically erect production equipment on top of the wellhead. But several companies are now developing sensors that can detect the mix of oil, gas and water near its point of entry deep within the well. "Smart" wells with such equipment will be able to separate water out of the well stream so that it never goes to the surface. Instead a pump, controlled by a computer in the drill pipe, will inject the wastewater below the oil level.

SEE *Figure 2 in color section.*

Wading in Deeper

Perhaps the oil industry's last great frontier is in deep water, in fields that lie 1,000 meters or more below the surface of the sea. Petroleum at such depths used to be beyond reach, but no longer. Remotely controlled robot submarines can now install on the seafloor the complex equipment needed to guard against blowouts, to regulate the flow of oil at the prevailing high pressures and to prevent natural gas from freezing and plugging pipelines. Subsea complexes will link clusters of horizontal wells. The collected oil will then be funneled both to tankers directly above and to existing platforms in shallower waters through long underwater pipelines. In just the next three years, such seafloor facilities are scheduled for construction in the Gulf of Mexico and off the shores of Norway, Brazil and West Africa.

More than deep water alone hinders the exploitation of offshore oil and gas fields. Large horizontal sheets of salt and basalt (an igneous rock) sometimes lie just underneath the seafloor in the deep waters of the continental margins. In conventional seismic surveys they scatter nearly all the sound energy so that oil fields below are hidden from view. But recently declassified U.S. Navy technology for measuring tiny variations in the force and direction of gravity, combined with ever-expanding supercomputer capabilities, now allows geophysicists to see under these blankets of salt or basalt.

Extracting oil from beneath the deep ocean is still enormously expensive, but innovation and necessity have led to a new wave of exploration in that realm. Already the 10 largest oil companies working in deep water have discovered new fields that will add 5 percent to their combined oil reserves, an increase not yet reflected in global reserve estimates.

SEE *Figure 3 in color section.*

The technology for oil exploration and production will continue to march forward in the 21st century. Although it is unlikely that these techniques will entirely eliminate the impending shortfall in the supply of crude oil, they will buy critical time for making an orderly transition to a world fueled by other energy sources.

—MARCH 1998

The Arctic Oil and Wildlife Refuge

The last great onshore oil field in America may lie beneath the nation's last great coastal wilderness preserve. Science can clarify the potential economics and the ecological risks of drilling into it.

W. WAYT GIBBS

Flying from Deadhorse, Alaska, west to Phillips Petroleum's new Alpine oil field, you can watch the evolution of oil development on the North Slope scroll below like a time-lapse film. At takeoff, the scene fills with the mammoth field where it all began: Prudhoe Bay, discovered in 1968 and uncorked in 1977 to send its oil down the Trans-Alaska Pipeline to the ice-free port at Valdez.

Climbing higher, the plane tracks feeder pipelines that zig westward to Kuparuk, second only to Prudhoe among the most oil-rich onshore fields yet found in North America. Like Prudhoe, Kuparuk has grown since its opening in 1981 into a scattershot of gravel well pads connected over 800 square miles by a web of roads and pipes to giant processing plants, camp buildings, vehicle lots and dark pits full of rock and mud drilled from the deep.

To the north, the artificial islands of Northstar and Endicott appear just offshore. And as the light descends onto the airstrip at Alpine, you fast-forward to the state of the art in petroleum engineering. Industry executives often cite this nearly roadless, 94-acre project as a model of environ-mentally and financially responsible oil development, proof that oil companies have learned how to coexist with delicate Arctic ecosystems.

Alpine is the newest and westernmost of the North Slope oil fields, but not for long. When its valves opened in November 2000, crude oil flowed the 50 miles back to Pump Station 1 near Deadhorse—as all oil produced on the slope must—via a new tributary to the pipeline system. By February, Alpine's production had already hit the plant's maximum output of almost 90,000 barrels a day. But the pipe to Deadhorse can carry much more.

It was built with the future in mind, and from Alpine the future of the hydrocarbon industry on the North Slope heads in three directions at once. It will continue westward, into the 23-million-acre National Petroleum Reserve–Alaska (NPR-A) on which Alpine borders. The federal government put four million acres up for lease in 1999, and exploration began last year. New fields there will deliver their oil through Alpine's pipe.

The future may lead southward as well. Soaring gas prices spurred North Slope companies last year to commit $75 million to plan a $10 billion natural gas pipeline that would open some 35 trillion cubic feet of untapped reserves to the lower American states by the end of the decade.

Beyond 2010, Phillips, BP and the other Alaskan oil producers look toward the east for new opportunities. Not 30 miles past Badami, the eastern terminus of the North Slope infrastructure, lie the coastal plain and tussock tundra of the so-called 1002 Area. It is named for the section of the Alaska National Interest Lands Conservation Act of 1980 that set aside 1.5 million acres of federal property in deference to geologists' guesses that the region entombs billions of barrels of oil and trillions of cubic feet of gas.

The same act placed the 1002 Area inside the 19-million-acre Arctic National Wildlife Refuge (ANWR), in deference to biologists' observations that the coastal plain provides a premium Arctic habitat: calving ground for the Porcupine caribou herd; nesting and staging wetlands for tundra swans and other migratory waterfowl; dens for polar bears and arctic foxes; and year-round forage for a small herd of musk oxen.

Congress thus instigated one of the longest-running environmental turf wars of the past century, and the darts have again begun to fly. On February 26, Senator Frank H. Murkowski of Alaska introduced S. 389, a bill that would open the 1002 Area to oil and gas exploration and production. The bill allows the Bureau of Land Management to restrict the activities to ensure that they "will result in no significant adverse effect on the fish and wildlife, their habitat, subsistence resources and the environment."

Can careful regulation prevent such effects? Or does even the most compact, high-tech, thoroughly monitored oil development pose an unac-

The Debate: Oil vs. Wildlife

- Senate bill S. 389 would open the coastal plain and foothills of the Arctic National Wildlife Refuge, the so-called 1002 Area, to oil development. A competing bill, S. 411, would designate the area as wilderness, prohibiting development.

- Geologists have used 1985 seismic data to estimate how much profitable oil and gas lies below the surface. But before any lease sale, oil companies would conduct new seismic surveys. That would leave a grid of visible scars in the vegetation of the plain but would have little or no effect on wildlife.

- Ice roads and exploration wells would follow. Fish and waterfowl may suffer if rivers and lakes are overdrained.

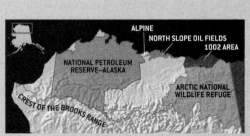

LAURIE GRACE

- A network of oil fields, processing plants and pipelines would extract the oil. A nearly roadless development may have little effect on the herd of 130,000 caribou that calves on the plain. Or it may displace the animals, affecting their nutrition, predation and birth rates, as well as long-term population growth.

ceptable risk to the largest American wildlife refuge remaining so close to its natural condition?

It is a mistake to ask scientists questions that force them to weigh the relative values of oil and wilderness. Some 245 biologists, not waiting to be asked, signed an open letter to President Bill Clinton last November urging him to bypass Congress and declare the area a wilderness, which would close it to development. In interviews with numerous Alaskan petroleum geologists, on the other hand, virtually all asserted that the oil industry could move in without causing more than cosmetic damage. In a fundamentally political dispute, scientists' opinions should carry no more weight than anyone else's.

Science and engineering should enter the debate over the fate of the Arctic refuge, however, not as a lobby but as a source of facts that all positions must accommodate. Thirty years of innovation have produced less disruptive ways of finding and removing the oil below the tundra. And 25 years of biology have quantified how those activities disturb the life on its surface. Before the public decides the question, it should have the clearest picture possible of what it might gain, what it might risk in the gamble—and what uncertainties are tucked into the word "might."

What Lies Beneath

At least eight separate groups of geologists have tried over the years to guess how much oil and gas sit below the 1002 Area in forms and places that would allow them to be recovered with current technology and at realistic prices. All eight teams relied on a single set of data from a seismic survey made in the winters of 1984 and 1985. Long rows of low-frequency microphones were set down on the snow to capture the echoes of sound-generating trucks up to a mile away as the sound waves bounced off rock layers at various depths. The string of microphones was moved, the process was

repeated, and 1,450 miles of cross-sectional snapshots were taken, covering the entire 1002 Area in a rough three-by-six-mile grid.

Turning those recordings into pictures of the subsurface and then inferring from the pictures which formations hold what quantity of oil is as much an art as it is a science. "The source rocks, trap formations [that hold the oil in place] and extent of migration all must be estimated based on analogies and prior experience," explains Mark D. Myers, director of the oil and gas division of Alaska's Department of Natural Resources. Wesley K. Wallace, a geologist at the University of Alaska, Fairbanks (UAF), ticks off more unknowns: "size of the formation, thickness, porosity—each has an error bar," sometimes a very large one, and even the size of the error bars is subjective.

No wonder, then, that the eight independent studies arrived at widely divergent estimates. Differences in their methods make it useless to compare them. But by all accounts, the best assessment to date is the latest one, led by Kenneth J. Bird of the U.S. Geological Survey (USGS). From 1996 to 1998 Bird and his teammates ran the old seismic data through new computer models. They gathered logs and rock samples from 41 wells drilled over the years near the borders of the refuge. They looked again at outcrops where oil-stained rock breaks through the permafrost and traveled to the adjacent mountains where some likely reservoir strata are uplifted and exposed. And they looked at the reflectance of vitrinite and the tracks made by radioactive nuclei in apatite found in the 1002 Area for clues to those minerals' temperature history, which matters because hydrocarbons turn into oil only when cooked just so.

The result is not one estimate but several, because the relevant figure is not how much oil is there but how much can be profitably recovered—and that depends in turn on the price of oil. Bird's group concluded that thorough exploration would

raised only over the course of several decades, following a classic bell-shaped curve. Industry insiders say that 10 years would probably pass between a decision to open the refuge to development and the first flow into the Alaskan pipeline. Environmental-impact studies and hearings would take two years, if the history of NPR-A is a guide. Companies would then have a year or two to do more intense seismic surveys and to prepare their bids on leases. Several years of exploration typically go into each discovery—after two years of drilling in NPR-A, for example, no strikes have been announced yet. Each permanent drilling site, processing facility and pipeline extension would have to clear more environmental analyses and hearings, and each would take two to three years to build.

An analysis by the U.S. Energy Information Administration (EIA) suggests that if the USGS estimate of seven bbo is correct, then the 1002 Area will generate fewer than 200,000 barrels a day for the first five years. The EIA also forecasts that American petroleum consumption, 19.5 million barrels a day last year, will rise to 23 million by 2010, with 66 percent of that amount imported. At its peak, probably no earlier than 2030, complete development of the coastal plain of the Arctic refuge would produce about one million barrels of oil a day. Flow from the 1002 Area would then meet something shy of 4 percent of the nation's daily demand for petroleum [see sidebar, "Facts: Forecasting the Flow"].

There's the Rub

Petroleum geologists know what they need to do to reduce the huge uncertainties in the USGS analysis. "The first thing a company would do is shoot a new 3-D seismic survey," Myers says. With gaps in the previous seismic data of up to six miles wide, "every prospect drilled on the slope this year would be invisible on that [1985] survey," he observes. This time "the grid would be much

most likely yield about seven billion barrels (bbo) of economically recoverable oil if North Slope prices remain above $24 a barrel, where they were in March. The estimate falls to about five bbo if oil prices slip to $18, and it plummets to a few hundred million barrels if prices drop to $12. Since 1991 the price of North Slope crude has fluctuated between $9 and $35, averaging $18 a barrel.

At seven bbo, the 1002 Area would hold about half as much profitable petroleum as Prudhoe Bay did in 1977. But as with Prudhoe, the oil could be

finer," with lines spaced about 1,100 feet apart, says Michael Faust, geosciences technology manager for Phillips in Anchorage. With new, high-resolution data, supercomputers could model the subsurface in three-dimensional detail.

The caravan of survey equipment, however, would appear much the same as before, Faust says: typically, eight vibrating and seven recording vehicles, accompanied by personnel carriers, mechanic trucks, mobile shop trucks, fuel tankers, an incinerator, plus a crew of 80 to 120 people and a camp train of 20 to 25 shipping containers on skis, pulled by several Caterpillar tractors on treads. The crew would leave in January and stay out through April, returning the next winter if necessary to cover the entire 1002 Area, 1,100 feet at a time. Each interested oil company or partnership would shoot its own complete survey, employing its own caravan.

That prospect worries Martha K. Raynolds, a UAF biologist. She and Janet C. Jorgenson, a botanist with the U.S. Fish and Wildlife Service in Fairbanks, have returned six times to monitor 200 patches of tundra that were randomly chosen for study as the last seismic vehicles passed over them 17 years ago. Ten percent still showed scuffing or reduced plant cover after 10 years, and 7 percent—about 100 miles of trail—had not recovered by 1998.

The problem, they say, is the terrain. The wide, low-pressure tires of the seismic trucks leave little trace on the flat, frozen, snow-covered grasslands around Prudhoe Bay and Alpine. Rubber treads on the tractors grip well enough. But east toward ANWR, the mountains march northward and the coast withdraws. That leaves the North Slope just 20 to 30 miles within the 1002 Area to attempt its typically gentle decline from rolling foothills to stream-crazed plateau to the ice-locked Beaufort Sea. Often it fails, and the tundra piles into hummocks. Winds clear the snow from their tops, exposing the dwarf willows and the standing dead

vegetation. Tires and skis crush the shrubs and compact the sedges. Rubber treads lose traction on slopes, are replaced with steel and inevitably dig in, Jorgenson says.

At breakup in May, permafrost below the compacted areas thaws early, deprived of its usual insulation. Pools form, some native plant species die, and visitors take over. Three quarters of the vegetative scars were still visible from the air a decade after the survey; many appear to be permanent. But no research suggests that the changes affect wildlife, both scientists say.

What Harm in Looking?

Seismic surveys generate clues, not discoveries. For petroleum geologists, truth emerges only from holes in the ground. Once the supercomputers have spit out their images, exploration teams would fan out across the frozen 1002 Area to drill wildcat wells. A mobile drill rig like the one at Alpine weighs 2.2 million pounds, so it is driven and parked on thick slabs of ice made by laying down six-inch-deep piles of ice chips and cementing them with water.

With lots of water, in fact—about a million gallons per mile of road. Around Prudhoe, tens of thousands of lakes ensure that liquid water is plentiful even when the air drops to −20 degrees Fahrenheit. Twelve years ago, however, a thorough search of the 1002 Area in April—when the ice hits its maximum thickness of seven feet—turned up only nine million gallons of liquid water sequestered in ice pockets along 237 miles of the major interior rivers. Steve Lyons, chief hydrologist for the refuge, found 255 lakes, ponds and puddles within the 1002 Area. Just 59 of those were deeper than seven feet, and only eight contained enough unfrozen water to build a mile or more of ice road. The largest basins lie in the Canning and Jago river deltas, and their bottom water is often brackish and potentially poisonous to vegetation.

Allow those few wet lakes to freeze through in winter, Lyons predicts, and next summer the waterfowl that pause in their migration to feed on invertebrates in the ponds will find fewer to eat. Draw too heavily from the spring-fed Canning, which runs free year-round, and the many kinds of fish that overwinter there may suffer, he warns.

"Water in ANWR could be a problem," says Thomas Manson, the environmental manager at Alpine, which treats and recycles its freshwater but still runs through 70,000 gallons every day. The trouble is not only quantity but also distribution: as a rule, water is drawn no farther than 10 miles from where it is needed, or else it freezes in the trucks on the way. Lyons admits that there may be technological solutions, such as a desalinization plant connected to a heated, elevated pipeline. But such measures would change the economics of the enterprise and thus the amount of oil recoverable.

(Wild) Life Goes On

Of course, if any oil is to be recovered, plants must be built. "Put four or five Alpine-size fields into ANWR with the processing facilities to support them, and you're talking about a few thousand acres of development," Myers says. "Clearly, some habitat will be damaged or destroyed. The question is: How will that modify the behavior of the animals?"

Theoretically, oil development could affect animals in many ways. Drillers no longer dump their cuttings and sewage and garbage into surface pits; these are now either burned or injected deep into wells. That greatly reduces the impact on foxes and bears. But there are other emissions. Alpine sees six to eight aircraft pass through every day, some as large as a C-130 Hercules. The scents of up to 700 workers and the noise of numerous trucks and two enormous turbines, big as the engines of a 747, constantly waft out over the tundra. A 10-foot gas flare shimmers atop a 100-foot stack. And three pipelines—two bringing seawater and diesel fuel in, one pumping crude out—fly to the horizon at just over the height of a caribou's antlers.

How the animal inhabitants of the 1002 Area would react to a collection of Alpine-style oil developments is a puzzle to which biologists have only pieces of a solution. Some wildlife does seem to have been displaced around the oil fields at Prudhoe and Kuparuk. Tundra swans, for example, tend to nest more than 650 feet from the roadways there, and caribou with calves typically hang back 2.5 miles or more.

Brad Griffith of UAF's Institute of Arctic Biology recently found two important patterns in the distribution since 1985 of the 130,000 caribou of the Porcupine herd, which arrives in the 1002 Area almost every year by June to bear and wean its young before departing for warmer climes by mid-July. The first pattern is a strong correlation of calf survival with the amount of high-protein food in the calving area. Second, caribou cows with newborns have consistently concentrated in the most rapidly greening areas (as measured by satellite) during lactation. Scott Wolfe, a graduate student of Griffith's, last year showed that the second pattern holds as well for the half of the Central Arctic herd that calves east of the Sagavanirktok River.

Across that river lie the big oil fields, and Wolfe found that from 1987 to 1995 the western half of the herd shifted its calving concentrations southward, away from the growing development and the richest forage. Ray Cameron, another Institute biologist, worries that that movement may affect the caribou numbers strongly enough to be perceptible above the normal fluctuations caused by weather, insect cycles and many other factors. It hasn't yet: at 27,000, the Central Arctic herd is five times as large as it was in 1978.

But in a 1995 study Cameron and others reported data showing that a 20-pound drop in the weight of the mother could lower calf survival by 20 percent and fertility by 30 percent. Cameron also tracked down radio-tagged cows and found that those that summered among the oil fields bore 23 percent fewer calves on average than their counterparts east of the river. But a critical link in this logical chain is missing: evidence that caribou, pushed off their preferred forage, don't get enough to eat.

Caribou in ANWR might suffer more than the Central Arctic herd has, because almost five times as many animals there forage in an area one fifth the size of the plain surrounding Prudhoe and Kuparuk. With fewer options, a larger fraction of the caribou cows may lose weight and bear fewer young. Oil fields could push more of them into the foothills, where calves are most likely to fall prey to eagles, wolves or bears. Griffith and his colleagues recently combined satellite imagery with caribou-calving and grizzly-bear-tracking data from the 1002 Area into a computer model. It predicts that pushing the caribou-calving concentration toward the foothills would reduce annual calf survival by 14 percent on average, Griffith says.

And Fish and Wildlife Service biologist Patricia Reynolds, who monitors the 250 musk oxen that live within the 1002 Area, points out that those animals survive the brutal winters on the plain primarily by moving little and conserving stored fat. If oil workers mine gravel from the riverbanks where they stand, the musk oxen will bolt, upsetting a precariously balanced energy budget and jeopardizing their young.

On the other hand, if the drill pads are served by short airstrips rather than long networks of roads, the caribou may fear them less and suffer little displacement. Wells no longer need be directly above the reservoir, so drill pads could be placed to avoid most nutritious cottongrass patches. Many of the musk oxen wear radio collars, so pains could be taken to avoid them.

All things considered, the wildlife would probably cope. The question is, could we? Science itself may have a vested interest in thwarting S. 389, suggests John W. Schoen, senior scientist with the Audubon Society in Anchorage. "If global climate is changing, its effects will be most magnified in northern latitudes, in places like the Arctic refuge," he argues. "How are we going to measure these subtle changes and sort out which are due to industrial development versus which are due to global climate change? One way is to protect some areas as experimental controls. The Arctic refuge would certainly serve as such a laboratory—if it remains intact."

In fact, the 1002 Area is already the centerpiece of a long and revealing experiment—a social and political experiment that may at last be approaching its conclusion. How the question is settled will reveal something about the American public's priorities, its patience, and its tolerance for risk.

—MAY 2001

CLIMATE REPAIR: CARBON CAPTURE

A Climate Repair Manual

Global warming is a reality. Innovation in energy technology and policy are sorely needed if we are to cope.

GARY STIX

Overview

- New reports pile up each month about the perils of climate change, including threats to marine life, increases in wildfires, even more virulent poison ivy.
- Implementing initiatives to stem global warming will prove more of a challenge than the Manhattan Project.
- Leading thinkers detail their ideas in the articles that follow for deploying energy technologies to decarbonize the planet.

Explorers attempted and mostly failed over the centuries to establish a pathway from the Atlantic to the Pacific through the icebound north, a quest often punctuated by starvation and scurvy. Yet within just 40 years, and maybe many fewer, an ascending thermometer will likely mean that the maritime dream of Sir Francis Drake and Captain James Cook will turn into an actual route of commerce that competes with the Panama Canal.

The term "glacial change" has taken on a meaning opposite to its common usage. Yet in reality, Arctic shipping lanes would count as one of the more benign effects of accelerated climate change. The repercussions of melting glaciers, disruptions in the Gulf Stream and record heat waves edge toward the apocalyptic: floods, pestilence, hurricanes, droughts—even itchier cases of poison ivy.

Month after month, reports mount of the deleterious effects of rising carbon levels. One recent study chronicled threats to coral and other marine organisms, another a big upswing in major wildfires in the western U.S. that have resulted because of warming.

The debate on global warming is over. Present levels of carbon dioxide—nearing 400 parts per million (ppm) in the earth's atmosphere—are higher than they have been at any time in the past 650,000 years and could easily surpass 500 ppm by the year 2050 without radical intervention.

The earth requires greenhouse gases, including water vapor, carbon dioxide and methane, to prevent some of the heat from the received solar radiation from escaping back into space, thus keeping the planet hospitable for protozoa, Shetland ponies and Lindsay Lohan. But too much of a good thing—in particular, carbon dioxide from SUVs and local coal-fired utilities—is causing a steady uptick in the thermometer. Almost all of the 20 hottest years on record have occurred since the 1980s.

No one knows exactly what will happen if things are left unchecked—the exact date when a polar ice sheet will complete a phase change from solid to liquid cannot be foreseen with precision, which is why the Bush administration and warming-skeptical public-interest groups still carry on about the uncertainties of climate change. But no climatologist wants to test what will arise if carbon dioxide levels drift much higher than 500 ppm.

A League of Rations

Preventing the transformation of the earth's atmosphere from greenhouse to unconstrained hothouse represents arguably the most imposing scientific and technical challenge that humanity has ever faced. Sustained marshaling of cross-border engineering and political resources over the course of a century or more to check the rise of carbon emissions makes a moon mission or a Manhattan Project appear comparatively straightforward.

Climate change compels a massive restructuring of the world's energy economy. Worries over fossil fuel supplies reach crisis proportions only when safeguarding the climate is taken into account. Even if oil production peaks soon—a debatable contention given Canada's oil sands, Venezuela's heavy oil and other reserves—coal and its derivatives could tide the earth over for more than a century. But fossil fuels, which account for 80 percent of the world's energy usage, become a liability if a global carbon budget has to be set.

Translation of scientific consensus on climate change into a consensus on what should be done about it carries the debate into the type of political minefield that has often undercut attempts at international governance since the League of Nations. The U.S. holds less than 5 percent of the world's population but produces nearly 25 percent of carbon emissions and has played the role of saboteur by failing to ratify the Kyoto Protocol and commit to reducing greenhouse gas emissions to 7 percent below 1990 levels.

Yet one of the main sticking points for the U.S.—the absence from that accord of a requirement that developing countries agree to firm emission limits—looms as even more of an obstacle as a successor agreement is contemplated to take effect when Kyoto expires in 2012. The torrid economic growth of China and India will elicit calls from industrial nations for restraints on emissions, which will again be met by even more adamant retorts that citizens of Shenzhen and Hyderabad should have the same opportunities to build their economies that those of Detroit and Frankfurt once did.

Kyoto may have been a necessary first step, if only because it lit up the pitted road that lies ahead. But stabilization of carbon emissions will require a more tangible blueprint for nurturing further economic growth while building a

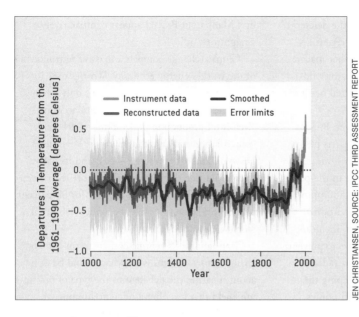

JEN CHRISTIANSEN, SOURCE: IPCC THIRD ASSESSMENT REPORT

This "hockey stick graph," from one of many studies showing a recent sharp increase in average temperatures, received criticism from warming skeptics, who questioned the underlying data. A report released in June by the National Research Council lends new credence to the sticklike trend line that traces an upward path of temperatures during the 20th century.

Greenhouse Effect

A prerequisite for life on earth, the greenhouse effect occurs when infrared radiation (heat) is retained within the atmosphere.

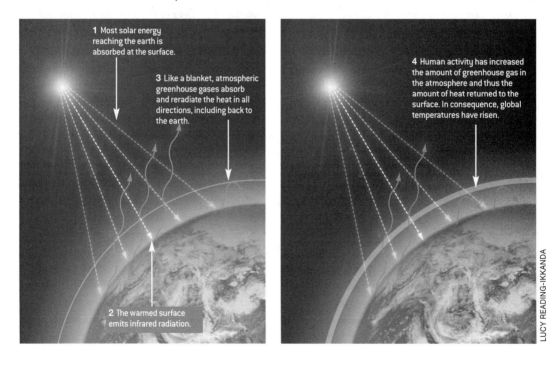

LUCY READING-IKKANDA

decarbonized energy infrastructure. An oil company's "Beyond Petroleum" slogans will not suffice.

Industry groups advocating nuclear power and clean coal have stepped forward to offer single-solution visions of clean energy. But too much devoted too early to any one technology could yield the wrong fix and derail momentum toward a sustainable agenda for decarbonization. Portfolio diversification underlies a plan laid out by Robert H. Socolow and Stephen W. Pacala ["A Plan to Keep Carbon in Check," page 52]. The two Princeton University professors describe how deployment of a basket of technologies and strategies can stabilize carbon emissions by midcentury.

Perhaps a solar-cell breakthrough will usher in the photovoltaic age, allowing both a steel plant and a cell phone user to derive all needed watts from a single source. But if that does not happen—and it probably won't—many technologies (biofuels, solar, hydrogen and nuclear) will be required to achieve a low-carbon energy supply. All these approaches are profiled by leading experts in this book, as are more radical ideas, such as solar power plants in outer space and fusion generators, which may come into play should today's seers prove myopic 50 years hence.

No More Business as Usual

Planning in 50- or 100-year increments is perhaps an impossible dream. The slim hope for keeping atmospheric carbon below 500 ppm hinges on aggressive programs of energy efficiency instituted by national governments. To go beyond what climate specialists call the "business as usual" scenario, the U.S. must follow Europe and even some of its own state governments in instituting new policies that affix a price on carbon—whether in the form of a tax on emissions or in a cap-and-trade system (emission allowances that are capped in aggregate at a certain level and then traded in open markets). These steps can furnish the breathing space to establish the defense-scale research programs needed to cultivate fossil fuel alternatives. The current federal policy vacuum has prompted a group of eastern states to develop their own cap-and-trade program under the banner of the Regional Greenhouse Gas Initiative.

Fifty-year time frames are planning horizons for futurists, not pragmatic policymakers. Maybe a miraculous new energy technology will simultaneously solve our energy and climate problems during that time, but another scenario is at least as likely: a perceived failure of Kyoto or international bickering over climate questions could foster the burning of abundant coal for electricity and synthetic fuels for transportation, both without meaningful checks on carbon emissions.

A steady chorus of skeptics continues to cast doubt on the massive amount of peer-reviewed scientific literature that forms the cornerstone for a consensus on global warming. "They call it pollution; we call it life," intones a Competitive Enterprise Institute advertisement on the merits of carbon dioxide. Uncertainties about the extent and pace of warming will undoubtedly persist. But the consequences of inaction could be worse than the feared economic damage that has bred overcaution. If we wait for an ice cap to vanish, it will simply be too late.

—SEPTEMBER 2006

Defusing the Global Warming Time Bomb

Global warming is real, and the consequences are potentially disastrous. Nevertheless, practical actions, which would also yield a cleaner, healthier atmosphere, could slow, and eventually stop, the process.

JAMES HANSEN

A paradox in the notion of human-made global warming became strikingly apparent to me one summer afternoon in 1976 on Jones Beach, Long Island. Arriving at midday, my wife, my son and I found a spot near the water to avoid the scorching hot sand. As the sun sank in the late afternoon, a brisk wind from the ocean whipped up whitecaps. My son and I had goose bumps as we ran along the foamy shoreline and watched the churning waves.

That same summer Andy Lacis and I, along with other colleagues at the NASA Goddard Institute for Space Studies, had estimated the effects of greenhouse gases on climate. It was well known by then that human-made greenhouse gases, especially carbon dioxide and chlorofluorocarbons (CFCs), were accumulating in the atmosphere. These gases are a climate "forcing," a perturbation imposed on the energy budget of the

Overview: Global Warming

- At present, our most accurate knowledge about climate sensitivity is based on data from the earth's history, and this evidence reveals that small forces, maintained long enough, can cause large climate change.

- Human-made forces, especially greenhouse gases, soot and other small particles, now exceed natural forces, and the world has begun to warm at a rate predicted by climate models.

- The stability of the great ice sheets on Greenland and Antarctica and the need to preserve global coastlines set a low limit on the global warming that will constitute "dangerous anthropogenic interference" with climate.

- Halting global warming requires urgent, unprecedented international cooperation, but the needed actions are feasible and have additional benefits for human health, agriculture and the environment.

planet. Like a blanket, they absorb infrared (heat) radiation that would otherwise escape from the earth's surface and atmosphere into space.

Our group had calculated that these human-made gases were heating the earth's surface at a rate of almost two watts per square meter. A miniature Christmas tree bulb dissipates about one watt, mostly in the form of heat. So it was as if humans had placed two of these tiny bulbs over every square meter of the earth's surface, burning night and day.

The paradox that this result presented was the contrast between the awesome forces of nature and the tiny lightbulbs. Surely their feeble heating could not command the wind and waves or smooth our goose bumps. Even their imperceptible heating of the ocean surface must be quickly dissipated to great depths, so it must take many years, perhaps centuries, for the ultimate surface warming to be achieved.

This seeming paradox has now been largely resolved through study of the history of the earth's climate, which reveals that small forces, maintained long enough, can cause large climate change. And, consistent with the historical evidence, the earth has begun to warm in recent decades at a rate predicted by climate models that take account of the atmospheric accumulation of human-made greenhouse gases. The warming is having noticeable impacts as glaciers are retreating worldwide, Arctic sea ice has thinned and spring comes about one week earlier than when I grew up in the 1950s.

Yet many issues remain unresolved. How much will our climate change in coming decades? What will be the practical consequences? What, if anything, should we do about it? The debate over these questions is highly charged because of the inherent economic stakes.

Objective analysis of global warming requires quantitative knowledge of three issues: the sensitivity of the climate system to forcings, the forcings that humans are introducing, and the time required for the climate to respond. All these issues can be studied with global climate models, which are numerical simulations on computers. But our most accurate knowledge about climate sensitivity, at least so far, is based on empirical data from the earth's history.

The Lessons of History

Over the past few million years the earth's climate has swung repeatedly between ice ages and warm interglacial periods. A 400,000-year record of temperature is preserved in the Antarctic ice sheet, which, except for coastal fringes, escaped melting even in the warmest interglacial periods. This record [see sidebar, "400,000 Years of Climate Change," page 26] suggests that the present interglacial period (the Holocene), now about 12,000 years old, is already long of tooth.

The natural millennial climate swings are associated with slow variations of the earth's orbit induced by the gravity of other planets, mainly Jupiter and Saturn (because they are so heavy) and Venus (because it comes so close). These perturbations hardly affect the annual mean solar energy striking the earth, but they alter the geographical and seasonal distribution of incoming solar energy, or insolation, as much as 20 percent. The insolation changes, over long periods, affect the building and melting of ice sheets.

Insolation and climate changes also affect uptake and release of carbon dioxide and methane by plants, soil and the ocean. Climatologists are still developing a quantitative understanding of the mechanisms by which the ocean and land release carbon dioxide and methane as the earth warms, but the paleoclimate data are already a gold mine of information. The most critical insight that the ice age climate swings provide is an empirical measure of climate sensitivity.

The composition of the ice age atmosphere is known precisely from air bubbles trapped as the

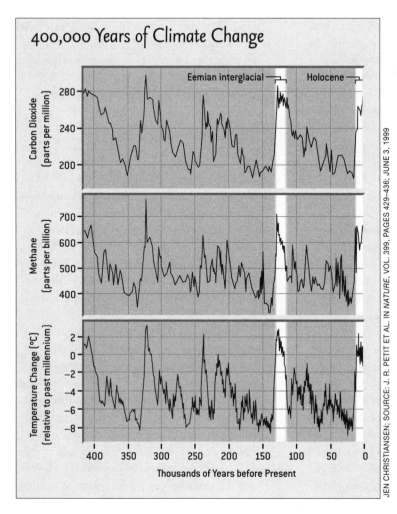

400,000 Years of Climate Change

Eemian interglacial — Holocene —

Carbon Dioxide (parts per million)

280
240
200

Methane (parts per billion)

700
600
500
400

Temperature Change [°C] (relative to past millennium)

2
0
-2
-4
-6
-8

400   350   300   250   200   150   100   50   0

Thousands of Years before Present

JEN CHRISTIANSEN; SOURCE: J. R. PETIT ET AL. IN *NATURE*, VOL. 399, PAGES 429–436; JUNE 3, 1999

Antarctic ice has preserved a 400,000-year record of temperature and of levels of carbon dioxide and methane in the atmosphere. Scientists study gases trapped in air bubbles in the ice—generally using ice cores extracted from the ice sheet and transported to a laboratory. The historical record provides us with two critical measures: Comparison of the current interglacial period (the Holocene) with the most recent ice age (20,000 years ago) gives an accurate measure of climate sensitivity to forcings. The temperature in the previous interglacial period (the Eemian), when sea level was several meters higher than today, defines an estimate of the warming that today's civilization would consider to be dangerous anthropogenic interference with climate.

Antarctic and Greenland ice sheets and numerous mountain glaciers built up from annual snowfall. Furthermore, the geographical distributions of the ice sheets, vegetation cover and coastlines during the ice age are well mapped. From these data we know that the change of climate forcing between the ice age and today was about 6.5 watts per square meter. This forcing maintains a global temperature change of 5 degrees Celsius (9 degrees Fahrenheit), implying a climate sensitivity of 0.75 ±0.25 degree C per watt per square meter. Climate models yield similar climate sensitivity.

The empirical result is more precise and reliable, however, because it includes all the processes operating in the real world, even those we have not yet been smart enough to include in the models.

The paleodata provide another important insight. Changes of the earth's orbit instigate climate change, but they operate by altering atmosphere and surface properties and thus the planetary energy balance. These atmosphere and surface properties are now influenced more by humans than by our planet's orbital variations.

Climate-Forcing Agents Today

The largest change of climate forcings in recent centuries is caused by human-made greenhouse gases. Greenhouse gases in the atmosphere absorb heat radiation rather than letting it escape into space. In effect, they make the proverbial blanket thicker, returning more heat toward the ground rather than letting it escape to space. The earth then is radiating less energy to space than it absorbs from the sun. This temporary planetary energy imbalance results in the earth's gradual warming.

Because of the large capacity of the oceans to absorb heat, it takes the earth about a century to approach a new balance—that is, for it to once again receive the same amount of energy from the sun that it radiates to space. And of course the balance is reset at a higher temperature. In the meantime, before it achieves this equilibrium, more forcings may be added.

The single most important human-made greenhouse gas is carbon dioxide, which comes mainly from burning fossil fuels (coal, oil and gas). Yet the combined effect of the other human-made gases is comparable. These other gases, especially tropospheric ozone and its precursors, including methane, are ingredients in smog that damage human health and agricultural productivity.

Aerosols (fine particles in the air) are the other main human-made climate forcing. Their effect is more complex. Some "white" aerosols, such as sulfates arising from sulfur in fossil fuels, are highly reflective and thus reduce solar heating of the earth; however, black carbon (soot), a product of incomplete combustion of fossil fuels, biofuels and outdoor biomass burning, absorbs sunlight and thus heats the atmosphere. This aerosol direct climate forcing is uncertain by at least 50 percent, in part because aerosol amounts are not well measured and in part because of their complexity.

Aerosols also cause an indirect climate forcing by altering the properties of clouds. The resulting brighter, longer-lived clouds reduce the amount of sunlight absorbed by the earth, so the indirect effect of aerosols is a negative forcing that causes cooling.

Other human-made climate forcings include replacement of forests by cropland. Forests are dark even with snow on the ground, so their removal reduces solar heating.

Natural forcings, such as volcanic eruptions and fluctuations of the sun's brightness, probably have little trend on a timescale of 1,000 years. But evidence of a small solar brightening over the past 150 years implies a climate forcing of a few tenths of a watt per square meter.

The net value of the forcings added since 1850 is 1.6 ±1.0 watts per square meter. Despite the large uncertainties, there is evidence that this estimated net forcing is approximately correct. One piece of evidence is the close agreement of observed global temperature during the past several decades with climate models driven by these forcings. More fundamentally, the observed heat gain by the world ocean in the past 50 years is consistent with the estimated net climate forcing.

Global Warming

Global average surface temperature has increased about 0.75 degree C during the period of extensive instrumental measurements, which began in the late 1800s. Most of the warming, about 0.5 degree C, occurred after 1950. The causes of observed warming can be investigated best for the past 50 years, because most climate forcings were observed then, especially since satellite measurements of the sun, stratospheric aerosols and ozone began in the 1970s. Furthermore, 70 percent of the anthropogenic increase of greenhouse gases occurred after 1950.

The most important quantity is the planetary energy imbalance [see sidebar, "Earth's Energy

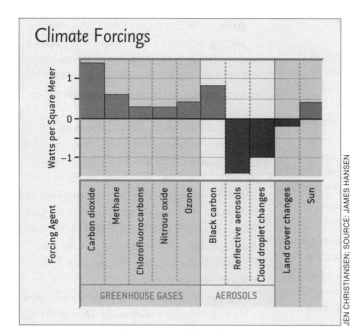

Climate Forcings

Watts per Square Meter

Forcing Agent

GREENHOUSE GASES   AEROSOLS

Carbon dioxide | Methane | Chlorofluorocarbons | Nitrous oxide | Ozone | Black carbon | Reflective aerosols | Cloud droplet changes | Land cover changes | Sun

A climate forcing is a mechanism that alters the global energy balance. A forcing can be natural—fluctuations in the earth's orbit, for example—or human-made, such as aerosols and greenhouse gases. Human-made climate forcings now dominate natural forcings. Carbon dioxide is the largest forcing, but air pollutants (black carbon, ozone, methane) together are comparable. (Aerosol effects are not known accurately.)

Imbalance," page 30]. This imbalance is a consequence of the long time that it takes the ocean to warm. We conclude that the earth is now out of balance by something between 0.5 and one watt per square meter—that much more solar radiation is being absorbed by the earth than is being emitted as heat to space. Even if atmospheric composition does not change further, the earth's surface will therefore eventually warm another 0.4 to 0.7 degree C.

Most of the energy imbalance has been heat going into the ocean. Sydney Levitus of the National Oceanic and Atmospheric Administration has analyzed ocean temperature changes of the past 50 years, finding that the world ocean heat content increased about 10 watt-years per square meter in the past 50 years. He also finds that the rate of ocean heat storage in recent years is consistent with our estimate that the earth is now out of energy balance by 0.5 to one watt per square meter. Note that the amount of heat required to melt enough ice to raise sea level one meter is about 12 watt-years (averaged over the planet),

energy that could be accumulated in 12 years if the planet is out of balance by one watt per square meter.

The agreement with observations, for both the modeled temperature change and ocean heat storage, leaves no doubt that observed global climate change is being driven by natural and anthropogenic forcings. The current rate of ocean heat storage is a critical planetary metric: it not only determines the amount of additional global warming already in the pipeline, but it also equals the reduction in climate forcings needed to stabilize the earth's present climate.

The Time Bomb

The goal of the United Nations Framework Convention on Climate Change, produced in Rio de Janeiro in 1989, is to stabilize atmospheric composition to "prevent dangerous anthropogenic interference with the climate system" and to achieve that goal in ways that do not disrupt the global economy. Defining the level of warming that constitutes "dangerous anthropogenic interference" is thus a crucial but difficult part of the problem.

The UN established an Intergovernmental Panel on Climate Change (IPCC) with responsibility for analysis of global warming. The IPCC has defined climate-forcing scenarios, used these for simulations of 21st-century climate and estimated the impact of temperature and precipitation changes on agriculture, natural ecosystems, wildlife and other matters. The IPCC estimates sea-level change as large as several tens of centimeters in 100 years, if global warming reaches several degrees Celsius. The group's calculated sea-level change is due mainly to thermal expansion of ocean water, with little change in ice-sheet volume.

These moderate climate effects, even with rapidly increasing greenhouse gases, leave the impression that we are not close to dangerous anthropogenic interference. I will argue, however, that we are much closer than is generally realized, and thus the emphasis should be on mitigating the changes rather than just adapting to them.

The dominant issue in global warming, in my opinion, is sea-level change and the question of how fast ice sheets can disintegrate. A large portion of the world's people live within a few meters of sea level, with trillions of dollars of infrastructure. The need to preserve global coastlines sets a low ceiling on the level of global warming that would constitute dangerous anthropogenic interference.

The history of the earth and the present human-made planetary energy imbalance together paint a disturbing picture about prospects for sea-level change. Data from the Antarctic temperature record show that the warming of the past 50 years has taken global temperature back to approximately the peak of the current interglacial (the Holocene). There is some additional warming in the pipeline that will take us about halfway to the highest global temperature level of the previous interglacial (the Eemian), which was warmer than

the Holocene, with sea level estimated to have been five to six meters higher. One additional watt per square meter of forcing, over and above that today, will take global temperature approximately to the maximum level of the Eemian.

The main issue is: How fast will ice sheets respond to global warming? The IPCC calculates only a slight change in the ice sheets in 100 years; however, the IPCC calculations include only the gradual effects of changes in snowfall, evaporation and melting. In the real world, ice-sheet disintegration is driven by highly nonlinear processes and feedbacks. The peak rate of deglaciation following the last ice age was a sustained rate of melting of more than 14,000 cubic kilometers a year—about one meter of sea-level rise every 20 years, which was maintained for several centuries. This period of most rapid melt coincided, as well as can be measured, with the time of most rapid warming.

Given the present unusual global warming rate on an already warm planet, we can anticipate that areas with summer melt and rain will expand over larger areas of Greenland and fringes of Antarctica. Rising sea level itself tends to lift marine ice shelves that buttress land ice, unhinging them from anchor points. As ice shelves break up, this accelerates movement of land ice to the ocean. Although building of glaciers is slow, once an ice sheet begins to collapse, its demise can be spectacularly rapid.

The human-induced planetary energy imbalance provides an ample supply of energy for melting ice. Furthermore, this energy source is supplemented by increased absorption of sunlight by ice sheets darkened by black-carbon aerosols and the positive feedback process as meltwater darkens the ice surface.

These considerations do not mean that we should expect large sea-level change in the next few years. Preconditioning of ice sheets for accelerated breakup may require a long time, perhaps many centuries. (The satellite ICESat, recently

Earth's Energy Imbalance

Reflected Energy from Atmosphere

Radiated Heat

Atmosphere

Heat Returned to Earth

Incoming Energy from Sun

Reflected Energy from Earth

TOTAL INCOMING SOLAR ENERGY	340 W/m²
TOTAL OUTGOING ENERGY	339 W/m²
REFLECTED ENERGY (from atmosphere and surface) 100 W/m² because of natural processes 1 W/m² because of human-made aerosols	101 W/m²
RADIATED HEAT (from land and ocean sinks) 240 W/m² because of natural processes −2 W/m² because of human-made greenhouse gases, which return heat to the surface	238 W/m²
NET RESULT	1 W/m²
1 W/m² of excess energy, which warms the oceans and melts glaciers and ice sheets	

IMBALANCE OVER PAST HALF-CENTURY

W/m² (change over time)

1.0   0   −1.0

1950 1960 1970 1980 1990 2000

Year

The earth's energy is balanced when the outgoing heat from the earth equals the incoming energy from the sun. At present the energy budget is not balanced (diagram and table). Human-made aerosols have increased reflection of sunlight by the earth, but this reflection is more than offset by the trapping of heat radiation by greenhouse gases. The excess energy—about one watt per square meter—warms the ocean and melts ice. The simulated planetary energy imbalance (graph) is confirmed by measurements of heat stored in the oceans. The planetary energy imbalance is a critical metric, in that it measures the net climate forcing and foretells future global warming already in the pipeline.

JEN CHRISTIANSEN; SOURCE: JAMES HANSEN

launched by NASA, may be able to detect early signs of accelerating ice-sheet breakup.) Yet I suspect that significant sea-level rise could begin much sooner if the planetary energy imbalance continues to increase. It seems clear that global warming beyond some limit will make a large sea-level change inevitable for future generations. And once large-scale ice-sheet breakup is under way, it will be impractical to stop. Dikes may protect limited regions, such as Manhattan and the Netherlands, but most of the global coastlines will be inundated.

I argue that the level of dangerous anthropogenic influence is likely to be set by the global temperature and planetary radiation imbalance at which substantial deglaciation becomes practically impossible to avoid. Based on the paleoclimate evidence, I suggest that the highest prudent level of additional global warming is not more than about one degree C. This means that additional climate forcing should not exceed about one watt per square meter.

Climate-Forcing Scenarios

The IPCC defines many climate-forcing scenarios for the 21st century based on multifarious "story lines" for population growth, economic development and energy sources. It estimates that added climate forcing in the next 50 years is one to three watts per square meter for carbon dioxide and two to four watts per square meter with other gases and aerosols included. Even the IPCC's minimum added forcing would cause dangerous anthropogenic interference with the climate system based on our criterion.

The IPCC scenarios may be unduly pessimistic, however. First, they ignore changes in emissions, some already under way, because of concerns about global warming. Second, they assume that true air pollution will continue to get worse, with ozone, methane and black carbon all greater in 2050 than in 2000. Third, they give short shrift to technology advances that can reduce emissions in the next 50 years.

An alternative way to define scenarios is to examine current trends of climate-forcing agents, to ask why they are changing as observed and to try to understand whether reasonable actions could encourage further changes in the growth rates.

The growth rate of the greenhouse gas climate forcing peaked in the early 1980s at almost 0.5 watt per square meter per decade but declined by the 1990s to about 0.3 watt per square meter per decade. The primary reason for the decline was reduced emissions of chlorofluorocarbons, whose production was phased out because of their destructive effect on stratospheric ozone.

The two most important greenhouse gases, with chlorofluorocarbons on the decline, are carbon dioxide and methane. The growth rate of carbon dioxide surged after World War II, flattened out from the mid-1970s to the mid-1990s and rose moderately in recent years to the current growth rate of about two parts per million per year. The methane growth rate has declined dramatically in the past 20 years, by at least two thirds.

These growth rates are related to the rate of global fossil fuel use. Fossil fuel emissions increased by more than 4 percent a year from the end of World War II until 1975 but subsequently by only about 1 percent a year. The change in fossil fuel growth rate occurred after the oil embargo and price increases of the 1970s, with subsequent emphasis on energy efficiency. Methane growth has also been affected by other factors, including changes in rice farming and increased efforts to capture methane at landfills and in mining operations.

If recent growth rates of these greenhouse gases continued, the added climate forcing in the next 50 years would be about 1.5 watts per square meter. To this must be added the change caused by other forcings, such as atmospheric ozone and aerosols. These forcings are not well monitored globally, but it is known that they are increasing in some countries while decreasing in others. Their net effect should be small, but it could add as much as 0.5 watt per square meter. Thus, if there is no slowing of emission rates, the human-made climate forcing could increase by two watts per square meter in the next 50 years.

This "current trends" growth rate of climate forcings is at the low end of the IPCC range of two to four watts per square meter. The IPCC four-watts-per-square-meter scenario requires 4

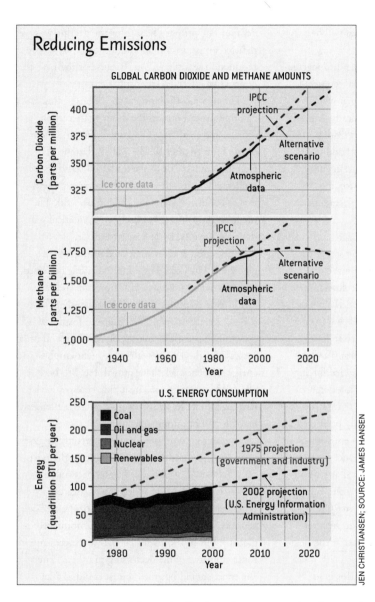

Reducing Emissions

GLOBAL CARBON DIOXIDE AND METHANE AMOUNTS

Carbon Dioxide (parts per million)

- IPCC projection
- Alternative scenario
- Atmospheric data
- Ice core data

Methane (parts per billion)

- IPCC projection
- Alternative scenario
- Atmospheric data
- Ice core data

Year: 1940, 1960, 1980, 2000, 2020

U.S. ENERGY CONSUMPTION

Energy (quadrillion BTU per year)

- ■ Coal
- ■ Oil and gas
- ■ Nuclear
- ■ Renewables

1975 projection (government and industry)

2002 projection (U.S. Energy Information Administration)

Year: 1980, 1990, 2000, 2010, 2020

Observed amounts of carbon dioxide and methane (top two graphs) fall below IPCC estimates, which have proved consistently pessimistic. Although the author's alternative scenario agrees better with observations, continuation on that path requires a gradual slowdown in carbon dioxide and methane emissions. Improvements in energy efficiency (bottom graph) have allowed energy use in the U.S. to fall below projections in recent decades, but more rapid efficiency gains are needed to achieve the carbon dioxide emissions of the alternative scenario, unless nuclear power and renewable energies grow substantially.

JEN CHRISTIANSEN; SOURCE: JAMES HANSEN

percent a year exponential growth of carbon dioxide emissions maintained for 50 years and large growth of air pollution; it is implausible.

Nevertheless, the "current trends" scenario is larger than the one-watt-per-square-meter level that I suggested as our current best estimate for the level of dangerous anthropogenic influence. This raises the question of whether there is a feasible scenario with still lower climate forcing.

A Brighter Future

I have developed a specific alternative scenario that keeps added climate forcing in the next 50 years at about one watt per square meter. It has two components: first, halt or reverse growth of air pollutants, specifically soot, atmospheric ozone and methane; second, keep average fossil fuel carbon dioxide emissions in the next 50 years about the same as today. The carbon dioxide and non–carbon dioxide portions of the scenario are equally important. I argue that they are feasible and at the same time protect human health and increase agricultural productivity.

In addressing air pollution, we should emphasize the constituents that contribute most to global warming. Methane offers a great opportunity. If human sources of methane are reduced, it may even be possible to get the atmospheric methane amount to decline, thus providing a cooling that would partially offset the carbon dioxide increase. Reductions of black-carbon aerosols would help counter the warming effect of reductions in sulfate aerosols. Atmospheric ozone precursors, besides methane, especially nitrogen oxides and volatile organic compounds, must be reduced to decrease low-level atmospheric ozone, the prime component of smog.

Actions needed to reduce methane, such as methane capture at landfills and at waste management facilities and during the mining of fossil fuels, have economic benefits that partially offset the costs. In some cases, methane's value as a fuel entirely pays for the cost of capture. Reducing black carbon would also have economic benefits, both in the decreased loss of life and work-years (minuscule soot particles carry toxic organic compounds and metals deep into lungs) and in increased agricultural productivity in certain parts of the world. Prime sources of black carbon are diesel fuels and biofuels (wood and cow dung, for example). These sources need to be dealt with for health reasons. Diesel could be burned more cleanly with improved technologies; however, there may be even better solutions, such as hydrogen fuel, which would eliminate ozone precursors as well as soot.

Improved energy efficiency and increased use of renewable energies might level carbon dioxide emissions in the near term. Long-term reduction of carbon dioxide emissions is a greater challenge, as energy use will continue to rise. Progress is needed across the board: continued efficiency improvements, more renewable energy and new technologies that produce little or no carbon dioxide or that capture and sequester it. Next-generation nuclear power, if acceptable to the public, could be an important contributor. There may be new technologies before 2050 that we have not imagined.

Observed global carbon dioxide and methane trends [see sidebar, "Reducing Emissions," page 32] for the past several years show that the real world is falling below all IPCC scenarios. It remains to be proved whether the smaller observed growth rates are a fluke, soon to return to IPCC rates, or are a meaningful difference. In contrast, the projections of my alternative scenario and the observed growth rates are in agreement. This is not surprising, because that scenario was defined with observations in mind. And in the three years since the alternative scenario was defined, observations have continued on that path. I am not suggesting, however, that the alternative scenario can be achieved without concerted efforts to reduce anthropogenic climate forcings.

How can I be optimistic if climate is closer to the level of dangerous anthropogenic interference than has been realized? If we compare the situation today with that of 10 to 15 years ago, we note that the main elements required to halt climate change have come into being with remarkable rapidity. I realize that it will not be easy to stabilize greenhouse gas concentrations, but I am optimistic because I expect that empirical evidence for climate change and its impacts will continue to accumulate and that this will influence the public, public-interest groups, industry and governments at various levels. The question is: Will we act soon enough?

—MARCH 2004

Capturing Greenhouse Gases

Sequestering carbon dioxide underground or in the deep ocean could help alleviate concerns about climate change.

HOWARD HERZOG, BALDUR ELIASSON AND OLAV KAARSTAD

The debate over climate change has shifted. Until very recently, scientists still deliberated whether human activity was altering the global climate. Specifically, was the release of greenhouse gases, which trap heat radiating from the earth's surface, to blame? With scientific evidence mounting in favor of the affirmative, the discussion is now turning to what steps society can take to protect our climate.

One solution almost certainly will not succeed: running out of fossil fuels—namely, coal, oil and natural gas. Morris Adelman, professor emeritus at the Massachusetts Institute of Technology and expert on the economics of oil and gas, has consistently made this point for 30 years. In the past century and a half, since the beginning of the industrial age, the concentration of carbon dioxide in the atmosphere has risen by almost one-third, from 280 to 370 parts per million (ppm)—primarily as a result of burning fossil fuels. In the 1990s, on average, humans discharged 1.5 ppm of carbon dioxide annually; with each passing year, the rate increased. Even though humans release other greenhouse gases, such as methane and nitrous oxide, experts project that carbon dioxide emissions will account for about two-thirds of potential global warming. As apprehension has grown regarding the possible hazards of a changing global climate, environmental groups, governments and certain industries have been trying to reduce the level of greenhouse gases in the atmosphere, often by promoting energy efficiency and alternative energy sources—for instance, wind or solar power.

Realistically, however, fossil fuels are cheap and plentiful and will be powering our cars, homes and factories well into the 21st century and possibly beyond. Worries about diminishing fuel supplies have surfaced periodically over the past 100 years, but continuing improvements in both oil exploration and production technology should keep the fuel flowing for decades to come. Furthermore, since the adoption of the first international treaty designed to stabilize greenhouse gas emissions, signed at the 1992 Earth Summit in Rio de Janeiro, the global demand for fossil fuels has actually increased. Today more than 85 percent of the world's commercial energy needs are supplied by fossil fuels. Although policies that promote energy efficiency and alternative energy sources are crucial to mitigating climate change, they are only one part of the solution.

Indeed, even if society were to cut back the use of fossil fuels today, the planet would still most likely experience significant repercussions as a result of past emissions. The climate's response time is slow, and carbon dioxide remains in the atmosphere for a century or more if left to nature's devices. Therefore, we must have a portfolio of technology options to adequately reduce the accelerating buildup of greenhouse gases. Significant research and development efforts are already exploring ways to improve energy efficiency and increase the use of fuels with no carbon content (renewable energy sources or nuclear power). But a third approach is attracting notice as people recognize that the first two options will simply not be sufficient: carbon sequestration, the idea of finding reservoirs where carbon dioxide can be stored rather than allowing it to build up in the atmosphere.

Our strategy may surprise some readers. Sequestering carbon is often connected to planting trees: trees (and vegetation in general) absorb carbon dioxide from the air as they grow and hold on to that carbon for their lifetime [see sidebar, "Plant a Tree," page 36]. Scientists estimate that, all together, plants currently retain about 600 gigatons of carbon, with another 1,600 gigatons in the soil.

Plants and soils could perhaps sequester another 100 gigatons or more of carbon, but additional sinks will be needed to meet the challenge of escalating greenhouse gas emissions. So during the past 10 years, the three of us have explored another possibility: capturing carbon dioxide from stationary sources—for example, a chemical factory or an electric power plant—and injecting it into the ocean or underground. We are not alone in our efforts but are part of a worldwide research community that includes the International Energy Agency (IEA) Greenhouse Gas Research and Development Program, as well as government and industry programs.

SEE *Figure 4 in color section.*

A New Approach in Norway

Sleipner offshore oil and natural gas field is in the middle of the North Sea, some 240 kilometers off the coast of Norway. Workers on one of the natural gas rigs there inject 20,000 tons of carbon dioxide each week into the pores of a sandstone layer 1,000 meters below the seabed. When the injection at Sleipner began in October 1996, it marked the first instance of carbon dioxide being stored in a geologic formation because of climate considerations.

How did this venture come about? One reservoir at Sleipner contains natural gas diluted with 9 percent carbon dioxide—too much for it to be attractive to customers, who generally accept no more than 2.5 percent. So, as is common practice at other natural gas fields around the world, an on-site chemical plant extracted the excess carbon dioxide. At any other installation, this carbon dioxide would simply be released to the atmosphere. But the owners of the Sleipner field—Statoil (where one of us, Kaarstad, works as a researcher), Exxon, Norsk Hydro and Elf—decided to sequester the greenhouse gas by first compressing it and then pumping it down a well into a 200-meter-thick sandstone layer, known as the Utsira Formation, which was originally filled with saltwater. The nearly one million tons of carbon dioxide sequestered at Sleipner last year may not seem large, but in the small country of Norway, it amounts to about 3 percent of total emissions to the atmosphere of this greenhouse gas.

The principal motivation for returning carbon to the ground at Sleipner was the Norwegian offshore carbon dioxide tax, which in 1996 amounted to $50 for every ton of the gas emitted (as of January 1, 2000, the tax was lowered to $38 per ton). The investment in the compression equipment and carbon dioxide well totaled around $80 million. In comparison, if the carbon dioxide had been emitted to the atmosphere, the companies would have owed about $50 million each year between 1996 and 1999. Thus, the savings paid off the investment in only a year and a half.

In other parts of the world, companies are planning similar projects. In the South China Sea, the Natuna field contains natural gas with nearly 71 percent carbon dioxide. Once this field has been developed commercially, the excess carbon dioxide will be sequestered. Other studies are investigating the possibility of storing captured carbon dioxide underground, including within liquefied natural gas installations at the Gorgon field on Australia's Northwest Shelf and the

Snøhvit ("Snow White") gas field in the Barents Sea off northern Norway, as well as the oil fields of Alaska's North Slope.

In all the projects now under way or in development, carbon dioxide must be captured for commercial reasons—for instance, to purify natu-

Plant a Tree

ANOTHER OPTION FOR STORING CARBON NEEDS ONLY SUN AND WATER

For over a decade, an organized carbon sequestration project has been under way in the deforested regions and farmlands of Guatemala. No underground pipes or pumping stations are required—just trees. As the plants grow, they absorb carbon dioxide from the atmosphere, which they store as carbon in the form of wood. Hoping to capitalize on this natural vehicle for sequestering carbon, companies and governments have initiated reforestation, afforestation (planting trees on land not previously forested) and agroforestry (integrating trees with agricultural crops) efforts as a way to meet obligations set forth in the Kyoto Protocol, the international environmental treaty on lowering greenhouse gas emissions.

In 1988, AES, a U.S.-based electrical company, pioneered the first forestry project designed to offset carbon dioxide emissions. At the time, AES was about to build a new coal-fired power plant in Connecticut, which was expected to release 52 million tons of carbon dioxide during its 40-year life span. Working in Guatemala with the World Resources Institute (WRI) and the relief organization CARE, AES

created community woodlots, introduced agroforestry practices and trained forest-fire brigades. According to WRI calculations, up to 58 million tons of carbon dioxide will be absorbed over the lifetime of the project. Currently more than a dozen such programs are under way on some four million hectares of forestland, including areas in the U.S., Norway, Brazil, Malaysia, Russia and Australia.

According to recent estimates, forests around the globe today store nearly one trillion tons of carbon. Scientists calculate that to balance current carbon dioxide emissions, people would have to plant new forests every year covering an area of land equivalent to the whole of India. Forestry projects are not a quick-fix solution, but they do offer many benefits, ranging from better habitats for wildlife to increased employment. Nevertheless, the potential for trees to serve as a reservoir for carbon is limited, and the approach has its drawbacks. Tree plantations drain native plant biodiversity and can disturb local communities, forcing them to relocate. As with many proposed solutions to climate change, trees will be effective only as one part of a global commitment to reduce greenhouse gas emissions.

—Diane Martindale

ral gas before it can be sold. The choice facing the companies involved is therefore between releasing the greenhouse gas to the atmosphere and storing it. They are not deciding whether to collect the carbon dioxide in the first place. We expect that more such companies needing to reduce carbon dioxide emissions will opt for sequestration in the future, but convincing other businesses to capture carbon dioxide emissions from large point sources such as electric power plants is more difficult because of the costs associated with carbon dioxide collection.

The Basics: Burying Carbon Dioxide

THE AUTHORS REVIEW CARBON SEQUESTRATION TECHNOLOGY

What is carbon sequestration? The idea is to store the greenhouse gas carbon dioxide in natural reservoirs rather than allowing it to build up in the atmosphere. Although sequestering carbon is often connected to planting trees, we are investigating the possibility of capturing carbon dioxide from stationary sources—an electric power plant, for example—and injecting it into the ocean or underground.

Where exactly will the carbon dioxide be stored? It can be pumped into underground geologic formations, such as unmineable coal beds, depleted oil or gas wells, or saline aquifers, in a process that is essentially the reverse of pumping oil up from below the earth's surface. Engineers are also looking into the possibility of bubbling carbon dioxide directly into the ocean at concentrations that will not affect the surrounding ecosystem and at depths that will ensure it remains in the ocean.

How will scientists make certain it is stored safely? Making sure carbon dioxide will be stored in a safe and environmentally sound manner is one of our primary goals. Memories of the 1986 Lake Nyos tragedy in Cameroon (in which a huge bubble of carbon dioxide erupted from the lake, suffocating some 1,700 people) raise the issue of safety, particularly for underwater storage. Yet the situation in the lake was entirely different from the scenario we envision for carbon sequestration in the ocean. A small lake simply cannot hold a large amount of carbon dioxide, so the Nyos eruption was inevitable. There are no such limitations in the oceans. In the case of underground storage, nature has demonstrated a safe track record: reservoirs such as the McElmo Dome in southwestern Colorado have held large quantities of carbon dioxide for centuries.

Are there any active carbon sequestration projects today? The Sleipner natural gas rig off the coast of Norway currently pumps carbon dioxide into a saline aquifer 1,000 meters below the seafloor. Although Sleipner is the only sequestration project driven solely by climatic change considerations, other commercial projects demonstrate the technology. More than a dozen power plants capture carbon dioxide from their flue gas, including the Shady Point, Oklahoma, plant built by the international engineering company ABB. And at over 65 oil wells in the U.S., companies inject the gas underground to enhance the efficiency of oil drilling.

Underground or Underwater

The technology for pumping carbon dioxide into the ground is actually well established—it is essentially the reverse of pumping oil and natural gas out of the ground. In fact, the practice is common at many oil fields today. Injecting carbon dioxide into an existing oil reservoir increases the mobility of the oil inside and thereby enhances the well's productivity. During 1998, U.S. oil field workers pumped a total of about 43 million tons of carbon dioxide into the ground at more than 65 enhanced oil recovery (EOR) projects. Yet this quantity adds up to comparatively little carbon sequestration. In contrast, geologic formations, including saline aquifer formations (such as that at Sleipner), unmineable coal beds, depleted oil or gas reservoirs, rock caverns and mined salt domes all around the world, can collectively hold hundreds if not thousands of gigatons of carbon.

Although geologic formations show great promise as storage sites, the largest potential reservoir for anthropogenic carbon dioxide is the deep ocean. Dissolved in its waters, the ocean holds an estimated 40,000 gigatons of carbon (compared with 750 gigatons in the atmosphere), but its capacity is much larger. Even if humans were to add to the ocean an amount of carbon dioxide equivalent to doubling the preindustrial atmospheric concentration of the gas, it would change the carbon content of the deep ocean by less than 2 percent. Indeed, slow-acting, natural processes will direct about 85 percent of present-day emissions into the oceans over hundreds of years. Our idea is to accelerate these events.

For ocean sequestration to be effective, the carbon dioxide must be injected into the sea below the thermocline—the layer of ocean between approximately 100 and 1,000 meters, in which water temperatures decrease dramatically with depth. The cooler, denser water below travels extremely slowly up through the thermocline. Therefore, the water beneath the thermocline may take centuries to mix with the surface waters, and any carbon dioxide below this boundary will be effectively trapped. In general, the deeper we inject the carbon dioxide, the longer it will take to reach the atmosphere.

Carbon dioxide can be introduced into seawater in two ways: dissolving it at moderate depths (from 1,000 to 2,000 meters) to form a dilute solution or injecting it below 3,000 meters to create what we call a carbon dioxide lake. The first strategy seeks to minimize local environmental effects by diluting the carbon dioxide, whereas the lake approach tries to maximize the length of time the carbon dioxide will reside in the ocean.

The concept of storing carbon dioxide in the ocean can be traced to a 1977 paper by Cesare Marchetti of the International Institute for Applied Systems Analysis in Laxenburg, Austria. Marchetti suggested that carbon dioxide could be piped into the waters of the Mediterranean Sea at Gibraltar, where it would naturally flow out into the Atlantic and be carried to the deep ocean. Even today building a pipe along the ocean floor to transport carbon dioxide to an appropriate depth remains one of the more realistic options for carbon sequestration. Other injection scenarios that have been suggested include dropping dry ice into the ocean from ships, introducing carbon dioxide at 1,000 meters through a pipe towed by a moving ship, and running a pipe down 3,000 meters or more to depressions in the seafloor.

Safe and Sound?

Despite the availability of the technology necessary to proceed with carbon storage in both terrestrial and oceanic reservoirs, we need to understand better what the consequences for the environment will be. Obviously, the process of storing carbon dioxide needs to be less damaging to the environment than the continued release of the greenhouse gas. In the case of underground storage, we must be sure to assess the long-term stability of any formation under consideration as a reservoir. The structural integrity of a site is important not only to ensure that the gas does not return to the atmosphere gradually but also because a sudden release of the carbon dioxide in a populated area could be catastrophic. Carbon dioxide is heavier than air, and a rapid, massive discharge of the gas would displace oxygen at the surface, suffocating people and wildlife. Fortunately, though, nature has stored carbon dioxide underground for millions of years in reservoirs such as McElmo Dome in southwestern Colorado, so we know there are ways to do it safely.

Ocean sequestration presents a different set of challenges. The leading concern is the repercussion it will have on the acidity of the ocean. Depending on the method of carbon dioxide release, the pH of seawater in the vicinity of an injection site could be between 5 and 7. (A pH of 7 is considered neutral; the pH of seawater is normally around 8.)

A large change in acidity could be harmful to organisms such as zooplankton, bacteria and bottom-dwelling creatures that cannot swim to less acidic waters. Research by one of us (Herzog) and MIT colleague E. Eric Adams, however, suggests that keeping the concentration of carbon dioxide dilute could minimize or even eliminate problems with acidity. For example, a dilution factor of one part per million yields a change in pH of less than 0.1. This reduced concentration could easily be achieved by releasing the carbon dioxide as small droplets from a pipe on the seafloor or on a moving ship.

Over the next several years, the scientific community will be conducting a number of experiments to assess how large amounts of carbon dioxide can be stored in a safe and environmentally sound manner. In the summer of 2001, for instance, a team of researchers from the U.S.,

Japan, Switzerland, Norway, Canada and Australia will begin a study off the Kona Coast of Hawaii to examine the technical feasibility and environmental effects of carbon storage in the ocean. (Two of us are participating in this project, Herzog as a member of the technical committee and Eliasson as a member of the steering committee.)

Our plan is to run a series of about 10 tests over a period of two weeks, involving the release of carbon dioxide at a depth of 800 meters. We will be monitoring the resulting plume and taking measurements, including the pH of the water and the amount of dissolved inorganic carbon. These data will allow us to refine computer models and thereby generalize the results of this experiment to predict environmental responses more accurately. We are also interested in what technical design works best to rapidly dilute the small droplets of carbon dioxide.

Money Matters

Along with questions of environmental safety and practicality, we must look at how much carbon sequestration will cost. Because electricity-generating power plants account for about one-third of all carbon dioxide released to the atmosphere worldwide and because such plants are large, concentrated sources of emissions, they provide a logical target for implementing carbon sequestration. Furthermore, such plants have had experience reducing pollutants in the past. (Notably, though, attention has primarily focused on controlling such contaminants as particulate matter, sulfur oxides, nitrogen oxides or even carbon monoxide—but not on carbon dioxide itself.)

Devices known as electrostatic precipitators, first introduced in the 1910s, helped to clean up the particles emitted from burning fossil fuels while raising the price of electricity only modestly. Today a modern power plant that includes state-of-the-art environmental cleanup equipment for particulates, sulfur oxides and nitrogen oxides costs up to 30 percent more to install than a plant without such equipment. This environmental equipment adds only between 0.1 and 0.5 of a cent per kilowatt-hour to the price of the electricity generated.

Because the exhaust gases of fossil-fueled power plants contain low concentrations of carbon dioxide (typically ranging from 3 to 15 percent), it would not be economical to funnel the entire exhaust stream into storage sites. The first step, therefore, should be to concentrate the carbon dioxide found in emissions. Unfortunately, with existing equipment this step turns out to be the most expensive. Thus, developing technology that lowers these costs is a major goal.

The most common method for separating carbon dioxide involves mixing a solution of dilute monoethanolamine (MEA) with the flue gases inside the absorption tower of a plant designed to capture the greenhouse gas. The carbon dioxide in the exhaust reacts with the MEA solution at room temperature to form a new, loosely bound compound. This compound is then heated in a second column, the stripping tower, to approximately 120 degrees C to release the carbon dioxide. The gaseous carbon dioxide product is then compressed, dried, chilled, liquefied and purified (if necessary); the liquid MEA solution is recycled. Currently, this technology works well, but it must become more energy efficient if it is to be applied to large-scale carbon sequestration. Today only a handful of power plants, including one built in Shady Point, Oklahoma, by ABB (where Eliasson serves as head of global change research), capture carbon dioxide from their flue gases. The carbon dioxide is then sold for commercial applications, such as freeze-drying chicken or carbonating beer and soda.

Another application for captured carbon dioxide offers a number of possible benefits. Methanol can be used as fuel even now. Generating this cleaner source of energy from captured carbon dioxide and hydrogen extracted from carbon-free sources would be more expensive than producing methanol from natural gas, as is currently done. But by reusing carbon dioxide—and by giving it a market value—this procedure ought to reduce overall emissions, provide an incentive to lower the costs of carbon dioxide–capture technology and help start a transition to more routine use of cleaner fuels.

Scientists, policymakers and the public must deal with the continuing importance of coal, oil and natural gas as a source of energy, even in a world constrained by concerns about climate change. The basic technology needed to use these fuels in a climate-friendly manner does exist. Current equipment for capturing carbon dioxide from power plants would raise the cost of generating electricity by 50 to 100 percent. But because sequestration does not affect the cost of electricity transmission and distribution (a significant portion of consumers' electricity bills), delivered prices will rise less, by about 30 to 50 percent. Research into better separation technologies should lead to lowered costs.

What needs to happen for carbon sequestration to become common practice? First, researchers need to verify the feasibility of the various proposed storage sites, in an open and publicly acceptable process. Second, we need leadership from industry and government to demonstrate these technologies on a large enough scale. Finally, we need improved technology to reduce costs associated with carbon dioxide separation from power plants. The Sleipner project has shown that carbon sequestration represents a realistic option to reduce carbon dioxide emissions when an economic incentive exists. During the past 100 years, our energy supply system has undergone revolutionary changes—from a stationary economy based on coal and steam to a mobile economy based on liquid fuels, gas and electricity. The changes over the next 100 years promise to be no less revolutionary.

—FEBRUARY 2000

A Breakthrough in Climate Change Policy?

David W. Keith and Edward A. Parson

As a result of human activities, the atmospheric concentration of carbon dioxide has increased by 31 percent over the past two centuries. According to business-as-usual projections, it will reach twice the preindustrial level before 2100. Although there is little doubt that this increase will noticeably transform the climate, substantial uncertainties remain about the magnitude, timing and regional patterns of climate change; even less is known about the ecological, economic and social consequences.

Despite these uncertainties, an international consensus has emerged regarding the importance of preventing runaway levels of carbon dioxide in the atmosphere. An effort to stabilize the concentration of carbon dioxide at even double its preindustrial level—generally considered the lowest plausible target—will require reducing global carbon dioxide emissions by about 50 percent from projected levels by 2050. Not surprisingly, such an extreme reduction will require a fundamental reorganization of global energy systems.

Most current assessments of greenhouse gas emissions assume that the reductions will be achieved through a mix of increasing energy efficiency and switching to nonfossil fuel alternative energy sources, such as solar, wind, biomass or nuclear. In the accompanying article, "Capturing Greenhouse Gases," the authors review a radically different approach: burning fossil fuels without releasing carbon dioxide to the atmosphere by separating the carbon emissions and burying them underground or in the deep ocean. We believe this approach—termed carbon management—has fundamental implications for the economics and politics of climate change.

Stabilizing the carbon dioxide concentration at 550 parts per million (ppm)—double the preindustrial level—is widely considered an ambitious target for emissions control. Yet this concentration will still cause substantial climate change. The resulting environmental problems, however, will most likely have only a small effect on the world's overall economic output; rich countries in particular should emerge relatively unscathed. But the results for specific regions will be more pronounced, with some places benefiting and others suffering. For instance, although parts of the northern U.S. may enjoy warmer winters, entire ecosystems, such as the southwestern mountain forests, alpine meadows and certain coastal forests, may disappear from the continental U.S. These likely consequences—and more important, the possibility of unanticipated changes—are compelling reasons to try to stabilize concentrations below 550 ppm, if it can be done at an acceptable cost.

At present, the cost of holding concentrations to even 550 ppm through conventional means appears high, both in dollars and in other environmental problems. All nonfossil fuel energy sources available today are expensive, and renewable sources have low power densities: they produce relatively little power for the amount of land required. Large-scale use of renewable energy could thereby harm our most precious environmental resource: land. Although technological advances should reduce the cost of renewables, little can be done to improve their power densities, which are intrinsic to the sources.

So must we conclude that reducing carbon emissions without causing other unacceptable environmental impacts will deliver a massive economic blow? Not necessarily. The crux of the cost problem is predicting how fast money-saving technical advances might develop in response to a carbon tax or some other form of regulation. Notably, most economic models used today to assess the cost of reducing emissions assume that innovation proceeds at its own pace and cannot be accelerated by policy. Under this assumption, delaying efforts to cut emissions makes sense because it will allow time to develop better technology that will lower the cost of reductions. Under the contrary assumption—which we regard as closer to the truth—innovation responds strongly to price and policy signals. In this case, early policy action on climate change is advantageous, because it would stimulate the innovations that reduce the cost of making large emission reductions.

Carbon management may be just such an innovation. Certain carbon management technologies are already available and appear to be significantly cheaper than renewables for generating electricity. To achieve deep reductions in greenhouse gas emissions, however, society must also start using carbon-free fuels, such as hydrogen, for transportation. Here, the relative advantage of carbon management over renewables is even greater than in producing electricity. Furthermore, these technologies offer one significant advantage over alternative energy sources: because they are more compatible with the existing energy infrastructure, we expect their costs to fall more quickly than those of renewables.

Carbon management weakens the link between burning fossil fuels and releasing greenhouse gases, making the world's economic dependence on fossil fuels more sustainable. This gives carbon management a crucial advantage: by reducing the threat to fossil fuel industries and fossil fuel–rich nations, carbon management may ease current political deadlocks. Stated bluntly, if society adopts carbon management widely, existing fossil fuel–dependent industries and nations may continue to operate profitably both in present energy markets and in new markets that develop around carbon management, making them more willing to tolerate policies that pursue substantial reduction of atmospheric emissions.

Environmentalists, however, are likely to find carbon management profoundly divisive for several reasons. Carbon sequestration is only as good as the reservoirs in which the carbon is stored. The unfortunate history of toxic and nuclear waste disposal has left many reasonable people skeptical of expert claims about the longevity of underground carbon disposal. As researchers assess the safety of proposed carbon reservoirs both underground and in the ocean, they must address such skepticism evenhandedly.

Perhaps even more disconcerting for environmentalists, though, is that carbon management collides with a deeply rooted belief that continued dependence on fossil fuels is an intrinsic problem, for which the only acceptable solution is renewable energy. Carbon management was first proposed as "geoengineering," a label it now shares with proposals to engineer the global climate, for example, by injecting aerosols into the stratosphere to reflect solar radiation and cool the earth's surface. Many environmentalists hold a reasonable distaste for large-scale technical fixes, arguing that it would be better to use energy sources that do not require such massive cleanup efforts.

Carbon management is a promising technology, but it remains unproved. And caution is certainly wise: the history of energy technologies is littered with options once touted as saviors that now play, at most, minor roles (for example, nuclear energy). Exploring the potential of either carbon management or renewable energy will require political and economic action now—that is, greater support for basic energy research and carbon taxes or equivalent policy measures that give firms incentives to develop and commercialize innovations that reduce emissions at a reasonable cost. It may be that carbon management will allow the world—at long last—to make deep cuts in carbon dioxide emissions at a politically acceptable cost. Indeed, for the next several decades, carbon management may be our best shot at protecting the global climate.

Can We Bury Global Warming?

Pumping carbon dioxide underground to avoid warming the atmosphere is feasible, but only if several key challenges can be met.

ROBERT H. SOCOLOW

Overview: Entombing CO2

- A strategy that combines the capture of carbon dioxide emissions from coal power plants and their subsequent injection into geologic formations for long-term storage could contribute significantly to slowing the rise of the atmospheric CO_2 concentration.

- Low-cost technologies for securing carbon dioxide at power plants and greater experience with CO_2 injection to avoid leakage to the surface are key to the success of large-scale CO_2 capture and storage projects.

- Fortunately, opportunities for affordable storage and capture efforts are plentiful. Carbon dioxide has economic value when it is used to boost crude oil recovery at mature fields. Natural gas purification and industrial hydrogen production yield CO_2 at low cost. Early projects that link these industries will enhance the practitioners' technical capabilities and will stimulate the development of regulations to govern CO_2 storage procedures.

When William Shakespeare took a breath, 280 molecules out of every million entering his lungs were carbon dioxide. Each time you draw breath today, 380 molecules per million are carbon dioxide. That portion climbs about two molecules every year.

No one knows the exact consequences of this upsurge in the atmosphere's carbon dioxide (CO_2) concentration nor the effects that lie ahead as more and more of the gas enters the air in the coming decades—humankind is running an uncontrolled experiment on the world. Scientists know that carbon dioxide is warming the atmosphere, which in turn is causing the sea level to rise, and that the CO_2 absorbed by the oceans is acidifying the water. But they are unsure of exactly how climate could alter across the globe, how fast the sea level might rise, what a more acidic ocean could mean, which ecological systems on land and in the sea would be most vulnerable to climate change and how these developments might affect human health and well-being. Our current course is bringing climate change upon ourselves faster than we can learn how severe the changes will be.

If slowing the rate of carbon dioxide buildup were easy, the world would be getting on with the job. If it were impossible, humanity would be working to adapt to the consequences. But reality lies in between. The task can be done with tools already at hand, albeit not necessarily easily, inexpensively or without controversy.

Were society to make reducing carbon dioxide emissions a priority—as I think it should to reduce the risks of environmental havoc in the future—we would need to pursue several strategies at once. We would concentrate on using energy more efficiently and on substituting noncarbon renewable or nuclear energy sources for fossil fuel (coal, oil and natural gas—the primary sources of man-made atmospheric carbon dioxide). And we would employ a method that is receiving increasing attention: capturing carbon dioxide and storing, or sequestering, it underground rather than releasing it into the atmosphere. Nothing says that CO_2 must be emitted into the air. The atmosphere has been our prime waste repository, because discharging exhaust up through smokestacks, tailpipes and chimneys is the simplest and least (immediately) costly thing to do. The good news is that the technology for capture and storage already exists and that the obstacles hindering implementation seem to be surmountable.

Carbon Dioxide Capture

The combustion of fossil fuels produces huge quantities of carbon dioxide. In principle, equipment could be installed to capture this gas wherever these hydrocarbons are burned, but some locations are better suited than others.

If you drive a car that gets 30 miles to the gallon and go 10,000 miles next year, you will need to buy 330 gallons—about a ton—of gasoline. Burning that much gasoline sends around three tons of carbon dioxide out the tailpipe. Although CO_2 could conceivably be caught before leaving the car and returned to the refueling station, no practical method seems likely to accomplish this task. On the other hand, it is easier to envision trapping the CO_2 output of a stationary coal-burning power plant.

It is little wonder, then, that today's capture-and-storage efforts focus on those power plants, the source of one-quarter of the world's carbon dioxide emissions. A new, large (1,000-megawatt-generating) coal-fired power plant produces six million tons of the gas annually (equivalent to the emissions of two million cars). The world's total output (roughly equivalent to the production of 1,000 large plants) could double during the next few decades as the U.S., China, India and many other countries construct new power-generating stations and replace old ones [see illustration on page 47]. As new coal facilities come online in the coming quarter of a century, they could be engineered to filter out the carbon dioxide that would otherwise fly up the smokestacks.

Today a power company planning to invest in a new coal plant can choose from two types of power systems, and a third is under development but not yet available. All three can be modified for carbon capture. Traditional coal-fired steam power plants burn coal fully in one step in air: the heat that is released converts water into high-pressure steam, which turns a steam turbine that generates electricity. In an unmodified version of this system—the workhorse of the coal power industry for the past century—a mixture of exhaust (or flue) gases exits a tall stack at atmospheric pressure after having its sulfur removed. Only about 15 percent of the flue gas is carbon dioxide; most of the remainder is nitrogen and water vapor. To adapt this technology for CO_2 capture, engineers could replace the smokestack with an absorption tower, in which the flue gases would come in contact with droplets of chemicals called amines that selectively absorb CO_2. In a second reaction column, known as a stripper tower, the amine liquid would be heated to release concentrated CO_2 and to regenerate the chemical absorber.

The other available coal power system, known as a coal gasification combined-cycle unit, first burns coal partially in the presence of oxygen in a gasification chamber to produce a "synthetic" gas, or syngas—primarily pressurized hydrogen and carbon monoxide. After removing sulfur compounds and other impurities, the plant combusts the syngas in air in a gas turbine—a modified jet engine—to make electricity. The heat in the exhaust gases leaving the gas turbine turns water into steam, which is piped into a steam turbine to generate additional power, and then the gas tur- bine exhaust flows out the stack. To capture carbon from such a facility, technicians add steam to the syngas to convert (or "shift") most of the carbon monoxide into carbon dioxide and hydrogen. The combined-cycle system next filters out the CO_2 before burning the remaining gas, now mostly hydrogen, to generate electricity in a gas turbine and a steam turbine.

The third coal power approach, called oxyfuel combustion, would perform all the burning in

Future Fossil-Fuel Power Plant

Consider a hypothetical town near a future 1,000-megawatt coal gasifica- tion power plant that has been sequestering carbon dioxide for 10 years. The town receives water from a shallow aquifer, unaffected by the CO_2 injection. The rail line transports coal to the plant, and the power lines carry away the electricity it generates.

Some 60 million tons of CO_2 have been captured during the plant's first 10 years of operation, and by now very large pancake- shaped deposits of CO_2 sit in the porous sub- terranean strata. The carbon dioxide was injected through horizontal wells into two deep brine (saltwater) formations, each located under impermeable caprock more than two kilometers below the surface. At seven-tenths the density of water, the high-pressure "supercritical" CO_2 occupies almost 90 million cubic meters. In both formations, 10 percent of the volume is pore space, and a third of the pores are filled with CO_2. Two-thirds of the injected gas has been pumped into the 40-meter-thick upper formation, and one-third has been sent into the 20-meter-thick lower formation. As a result, the total (horizontal) area of porous rock soaked with supercritical carbon dioxide in each forma- tion is about 40 square kilometers.

Note that the horizontal and vertical scales depicted differ. The depth of each injection well and the length of their horizontal extensions are really about equal in length, around two kilometers. Nor are the building structures to scale.

Technicians at a seismic monitoring sta- tion keep track of the CO_2 locations by beam- ing sound waves into the ground. During the power station's initial decade of operation, util- ity managers learned many details about the local geology by observing how the CO_2 spread through the area. This information will help them decide whether to continue injecting the plant emissions down the same wells, to bore new holes into the same formations, or to switch to alternative underground formations.

(SEE) *Figure 5 in color section.*

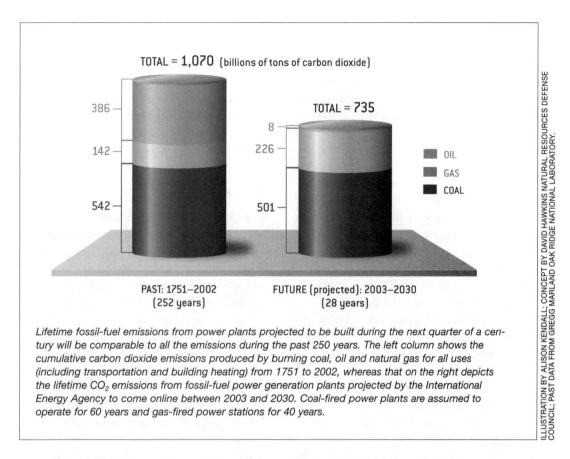

TOTAL = **1,070** (billions of tons of carbon dioxide)

386

142

542

TOTAL = **735**

8

226

501

OIL
GAS
COAL

PAST: 1751–2002
(252 years)

FUTURE (projected): 2003–2030
(28 years)

Lifetime fossil-fuel emissions from power plants projected to be built during the next quarter of a century will be comparable to all the emissions during the past 250 years. The left column shows the cumulative carbon dioxide emissions produced by burning coal, oil and natural gas for all uses (including transportation and building heating) from 1751 to 2002, whereas that on the right depicts the lifetime CO_2 emissions from fossil-fuel power generation plants projected by the International Energy Agency to come online between 2003 and 2030. Coal-fired power plants are assumed to operate for 60 years and gas-fired power stations for 40 years.

oxygen instead of air. One version would modify single-step combustion by burning coal in oxygen, yielding a fuel gas with no nitrogen, only CO_2 and water vapor, which are easy to separate. A second version would modify the coal gasification combined-cycle system by using oxygen, rather than air, at the gas turbine to burn the carbon monoxide and hydrogen mixture that has exited the gasifier. This arrangement skips the shift reaction and would again produce only CO_2 and water vapor. Structural materials do not yet exist, though, that can withstand the higher temperatures that are created by combustion in oxygen rather than in air. Engineers are exploring whether reducing the process temperature by recirculating

the combustion exhaust will provide a way around these materials constraints.

Tough Decisions

Modification for carbon dioxide capture not only adds complexity and expense directly but also cuts the efficiency of extracting energy from the fuel. In other words, safely securing the carbon byproducts means mining and burning more coal. These costs may be partially offset if the plant can filter out gaseous sulfur simultaneously and store it with the CO_2, thus avoiding some of the considerable expense of sulfur treatment.

Utility executives want to maximize profits over the entire life of the plant, probably 60 years or more, so they must estimate the expense of complying not only with today's environmental rules but also with future regulations. The managers know that the extra costs for CO_2 capture are likely to be substantially lower for coal gasification combined-cycle plants than for traditional plants. Removing carbon dioxide at high pressures, as occurs in a syngas operation, costs less because smaller equipment can be employed. But they also know that only a few demonstration gasification plants are running today, so that opting for gasification will require spending extra on backup equipment to ensure reliability. Hence, if the management bets on not having to pay for CO_2 emissions until late in the life of its new plant, it will probably choose a traditional coal plant, although perhaps one with the potential to be modified later for carbon capture. If, however, it believes that government directives to capture CO_2 are on their way within a decade or so, it may select a coal gasification plant.

To get a feel for the economic pressures the extra cost of carbon sequestration would place on the coal producer, the power plant operator and the homeowner who consumes the electricity, it helps to choose a reasonable cost estimate and then gauge the effects. Experts calculate that the total additional expense of capturing and storing a ton of carbon dioxide at a coal gasification combined-cycle plant will be about $25. (In fact, it may be twice that much for a traditional steam plant using today's technology. In both cases, it will cost less when new technology is available.)

The coal producer, the power plant operator and the homeowner will perceive that $25 cost increase quite differently. A coal producer would see a charge of about $60 per ton of coal for capturing and storing the coal's carbon, roughly tripling the cost of coal delivered to an electric utility customer. The owner of a new coal power plant would face a 50 percent rise in the cost of

power the coal plant puts on the grid, about two cents per kilowatt-hour (kWh) on top of a base cost of around four cents per kWh. The homeowner buying only coal-based electricity, who now pays an average of about 10 cents per kWh, would experience one-fifth higher electricity costs (provided that the extra two cents per kWh cost for capture and storage is passed on without increases in the charges for transmission and distribution).

First and Future Steps

Rather than waiting for the construction of new coal-fired power plants to begin carbon dioxide capture and storage, business leaders are starting the process at existing facilities that produce hydrogen for industry or purify natural gas (methane) for heating and power generation. These operations currently generate concentrated streams of CO_2. Industrial hydrogen production processes, located at oil refineries and ammonia plants, remove carbon dioxide from a high-pressure mix of CO_2 and hydrogen, leaving behind carbon dioxide that is released skyward. Natural gas purification plants must remove CO_2 because the methane is heading for a liquefied natural gas tanker and must be kept free of cold, solid carbon dioxide (dry ice) that could clog the system or because the CO_2 concentration is too high (above 3 percent) to be allowed on the natural gas distribution grid.

Many carbon dioxide capture projects using these sources are now under consideration throughout the oil and gas industry. Hydrogen production and natural gas purification are the initial stepping-stones to full-scale carbon capture at power plants; worldwide about 5 percent as much carbon dioxide is produced in these two industries as in electric power generation.

In response to the growing demand for imported oil to fuel vehicles, some nations, such as China, are turning to coal to serve as a feedstock for synthetic fuels that substitute for gasoline and diesel fuel. From a climate change perspective, this is a step backward. Burning a coal-based synthetic fuel rather than gasoline to drive a set distance releases approximately double the carbon dioxide, when one takes into account both tailpipe and synfuels plant emissions. In synthetic fuels production from coal, only about half the carbon in the coal ends up in the fuel, and the other half is emitted at the plant. Engineers could modify the design of a coal synfuels plant to capture the plant's CO_2 emissions. At some point in the future, cars could run on electricity or carbon-free hydrogen extracted from coal at facilities where CO_2 is captured.

Electricity can also be made from biomass fuels, a term for commercial fuels derived from plant-based materials: agricultural crops and residues, timber and paper industry waste and landfill gas. If the fossil fuels used in harvesting and processing are ignored, the exchanges between the atmosphere and the land balance because the quantity of carbon dioxide released by a traditional biomass power plant nearly equals that removed from the

Alternative CO2 Storage Schemes

Captured carbon dioxide might be stored not only in depleted oil and gas reservoirs and subterranean brine formations but also in minerals that form carbonate compounds, in coal seams and in the deep ocean.

Minerals that can become carbonates could potentially sequester even more carbon dioxide on the earth's surface than brine formations could store underground. The magnesium oxide in two abundant iron-magnesium minerals, serpentine and olivine, combines with CO_2 to produce highly stable magnesium carbonate. The big challenge is to get CO_2 to react quickly with bulk quantities of these rocks, perhaps by grinding them into fine powders to increase the surface area at which the chemical reactions occur.

The pore surfaces within coal formations adsorb methane. During mining, some of this methane can be released, too often causing underground explosions and, consequently, the deaths of miners. Pressurized carbon dioxide could be introduced into unexploited coal seams where it would replace the adsorbed methane, which could then be recovered and sold as fuel.

Ocean injection of carbon dioxide is controversial. Advocates of storage in the deep ocean point out that atmospheric CO_2 passes continuously into the ocean surface, as the air and ocean system seeks chemical equilibrium. Slowing the increase of CO_2 levels in the air will reduce the amount dissolving into the surface water. Thus, deep-ocean injection would shift some CO_2 from the surface waters to the lowest layers, reducing environmental impacts near the surface, where most marine life is found. Opponents of ocean storage cite international law that protects the oceans from certain kinds of industrial uses and the difficulties of monitoring carbon dioxide transport after injection. In many parts of the world, opponents tap into a strong cultural preference for leaving the oceans alone.

atmosphere by photosynthesis when the plants grew. But biomass power can do better: if carbon capture equipment was added to these facilities and the harvested biomass vegetation was replanted, the net result would be to scrub the air of CO_2. Unfortunately, the low efficiency of photosynthesis limits the opportunity for atmospheric scrubbing because of the need for large land areas to grow the trees or crops. Future technologies may change that, however. More efficient carbon dioxide removal by green plants and direct capture of CO_2 from the air (accomplished, for example, by flowing air over a chemical absorber) may become feasible at some point.

Carbon Dioxide Storage

Carbon capture is just half the job, of course. When an electric utility builds a 1,000-megawatt coal plant designed to trap CO_2, it needs to have somewhere to stash securely the six million tons of the gas the facility will generate every year for its entire life. Researchers believe that the best destinations in most cases will be underground formations of sedimentary rock loaded with pores now filled with brine (salty water). To be suitable, the sites typically would lie far below any source of drinking water, at least 800 meters under the surface. At 800 meters, the ambient pressure is 80 times that of the atmosphere, high enough that the pressurized injected CO_2 is in a "supercritical" phase—one that is nearly as dense as the brine it replaces in geologic formations. Sometimes crude oil or natural gas will also be found in the brine formations, having invaded the brine millions of years ago.

The quantities of carbon dioxide sent below ground can be expressed in "barrels," the standard 42-gallon unit of volume employed by the petroleum industry. Each year at a 1,000-megawatt coal plant modified for carbon capture, about 50 million barrels of supercritical carbon dioxide would

be secured—about 100,000 barrels a day. After 60 years of operation, about three billion barrels (half a cubic kilometer) would be sequestered below the surface. An oil field with a capacity to produce three billion barrels is six times the size of the smallest of what the industry calls "giant" fields, of which some 500 exist. This means that each large modified coal plant would need to be associated with a "giant" CO_2 storage reservoir. About two thirds of the 1,000 billion barrels of oil the world has produced to date has come from these giant oil fields, so the industry already has a good deal of experience with the scale of the operations needed for carbon storage.

Many of the first sequestration sites will be those that are established because they can turn a profit. Among these are old oil fields into which carbon dioxide can be injected to boost the production of crude. This so-called enhanced oil recovery process takes advantage of the fact that pressurized CO_2 is chemically and physically suited to displacing hard-to-get oil left behind in the pores of the geologic strata after the first stages of production. In this process, compressors drive CO_2 into the oil remaining in the deposits, where chemical reactions result in modified crude oil that moves more easily through the porous rock toward production wells. In particular, CO_2 lowers crude oil's interfacial tension—a form of surface tension that determines the amount of friction between the oil and rock. Thus, carbon dioxide injects new life into old fields.

In response to British government encouragement of carbon dioxide capture and storage efforts, oil companies are proposing novel capture projects at natural gas power plants that are coupled with enhanced oil recovery ventures at fields underneath the North Sea. In the U.S., operators of these kinds of fields can make money today while paying $10 to $20 per ton for carbon dioxide delivered to the well. If oil prices continue to rise, however, the value of injected CO_2 will probably go up because its use enables the production of a more valuable commodity. This market development could lead to a dramatic expansion of carbon dioxide capture projects.

Carbon sequestration in oil and gas fields will most likely proceed side by side with storage in ordinary brine formations, because the latter structures are far more common. Geologists expect to find enough natural storage capacity to accommodate much of the carbon dioxide that could be captured from fossil fuels burned in the 21st century.

Storage Risks

Two classes of risk must be addressed for every candidate storage reservoir: gradual and sudden leakage. Gradual release of carbon dioxide merely returns some of the greenhouse gas to the air. Rapid escape of large amounts, in contrast, could have worse consequences than not storing it at all. For a storage operation to earn a license, regulators will have to be satisfied that gradual leakage can occur only at a very slow rate and that sudden leakage is extremely unlikely.

Although carbon dioxide is usually harmless, a large, rapid release of the gas is worrisome because high concentrations can kill. Planners are well aware of the terrible natural disaster that occurred in 1986 at Lake Nyos in Cameroon: carbon dioxide of volcanic origin slowly seeped into the bottom of the lake, which sits in a crater. One night an abrupt overturning of the lake bed let loose between 100,000 and 300,000 tons of CO_2 in a few hours. The gas, which is heavier than air, flowed down through two valleys, asphyxiating 1,700 nearby villagers and thousands of cattle. Scientists are studying this tragedy to ensure that no similar man-made event will ever take place. Regulators of storage permits will want assurance that leaks cannot migrate to belowground confined spaces that are vulnerable to sudden release.

Gradual leaks may pose little danger to life, but they could still defeat the climate goals of sequestration. Therefore, researchers are examining the conditions likely to result in slow seepage. Carbon dioxide, which is buoyant in brine, will rise until it hits an impermeable geologic layer (caprock) and can ascend no farther.

Carbon dioxide in a porous formation is like hundreds of helium balloons, and the solid caprock above is like a circus tent. A balloon may escape if the tent has a tear in it or if its surface is tilted to allow a path for the balloon to move sideways and up. Geologists will have to search for faults in the caprock that could allow escape as well as determine the amount of injection pressure that could fracture it. They will also evaluate the very slow horizontal flow of the carbon dioxide outward from the injection locations. Often the sedimentary formations are huge, thin pancakes. If carbon dioxide is injected near the middle of a pancake with a slight tilt, it may not reach the edge for tens of thousands of years. By then, researchers believe, most of the gas will have dissolved in the brine or have been trapped in the pores.

Even if the geology is favorable, using storage formations where there are old wells may be problematic. More than a million wells have been drilled in Texas, for example, and many of them were filled with cement and abandoned. Engineers are worried that CO_2-laden brine, which is acidic, could find its way from an injection well to an abandoned well and thereupon corrode the cement plug and leak to the surface. To find out, some researchers are now exposing cement to brine in the laboratory and sampling old cements from wells. This kind of failure is less likely in carbonate formations than in sandstone ones; the former reduce the destructive potency of the brine.

The world's governments must soon decide how long storage should be maintained. Environmental ethics and traditional economics give different answers. Following a strict environmental ethic that seeks to minimize the impact of today's activities on future generations, authorities might, for instance, refuse to certify a storage project estimated to retain carbon dioxide for only 200 years. Guided instead by traditional economics, they might approve the same project on the grounds that two centuries from now a smarter world will have invented superior carbon disposal technology.

The next few years will be critical for the development of carbon dioxide capture-and-storage methods, as policies evolve that help to make CO_2-emission reduction profitable and as licensing of storage sites gets under way. In conjunction with significant investments in improved energy efficiency, renewable energy sources and, possibly, nuclear energy, commitments to capture and storage can reduce the risks of global warming.

—JULY 2005

A Plan to Keep Carbon in Check

Getting a grip on greenhouse gases is daunting but doable. The technologies already exist. But there is no time to lose.

ROBERT H. SOCOLOW AND STEPHEN W. PACALA

Overview

- Humanity can emit only so much carbon dioxide into the atmosphere before the climate enters a state unknown in recent geologic history and goes haywire. Climate scientists typically see the risks growing rapidly as CO_2 levels approach a doubling of their pre-18th-century value.
- To make the problem manageable, the required reduction in emissions can be broken down into "wedges"—an incremental reduction of a size that matches available technology.

Retreating glaciers, stronger hurricanes, hotter summers, thinner polar bears: the ominous harbingers of global warming are driving companies and governments to work toward an unprecedented change in the historical pattern of fossil fuel use. Faster and faster, year after year for two centuries, human beings have been transferring carbon to the atmosphere from below the surface of the earth. Today the world's coal, oil and natural gas industries dig up and pump out about seven billion tons of carbon a year, and society burns nearly all of it, releasing carbon dioxide (CO_2). Ever more people are convinced that prudence dictates a reversal of the present course of rising CO_2 emissions.

The boundary separating the truly dangerous consequences of emissions from the merely unwise is probably located near (but below) a doubling of the concentration of CO_2 that was in the atmosphere in the 18th century, before the Industrial Revolution began. Every increase in concentration carries new risks, but avoiding that danger zone would reduce the likelihood of triggering major, irreversible climate changes, such as the disappearance of the Greenland ice cap. Two years ago the two of us provided a simple framework to relate future CO_2 emissions to this goal.

We contrasted two 50-year futures. In one future, the emissions rate continues to grow at the pace of the past 30 years for the next 50 years, reaching 14 billion tons of carbon a year in 2056. (Higher or lower rates are, of course, plausible.) At that point, a tripling of preindustrial carbon concentrations would be very difficult to avoid, even with concerted efforts to decarbonize the world's energy systems over the following 100 years. In the other future, emissions are frozen at the present value of seven billion tons a year for the next 50 years and then reduced by about half over the following 50 years. In this way, a doubling of CO_2 levels can be avoided. The difference between these 50-year emission paths—one ramping up and one flattening out—we called the stabilization triangle [see sidebar, "Managing the Climate Problem," page 54].

To hold global emissions constant while the world's economy continues to grow is a daunting task. Over the past 30 years, as the gross world product of goods and services grew at close to 3 percent a year on average, carbon emissions rose half as fast. Thus, the ratio of emissions to dollars of gross world product, known as the carbon intensity of the global economy, fell about 1.5 percent a year. For global emissions to be the same in 2056 as today, the carbon intensity will need to fall not half as fast but fully as fast as the global economy grows.

Two long-term trends are certain to continue and will help. First, as societies get richer, the serv-ices sector—education, health, leisure, banking and so on—grows in importance relative to energy-intensive activities, such as steel production. All by itself, this shift lowers the carbon intensity of an economy.

Second, deeply ingrained in the patterns of technology evolution is the substitution of cleverness for energy. Hundreds of power plants are not needed today because the world has invested in much more efficient refrigerators, air conditioners and motors than were available two decades ago. Hundreds of oil and gas fields have been developed more slowly because aircraft engines consume less fuel and the windows in gas-heated homes leak less heat.

The task of holding global emissions constant would be out of reach were it not for the fact that all the driving and flying in 2056 will be in vehicles not yet designed, most of the buildings that will be around then are not yet built, the locations of many of the communities that will contain these buildings and determine their inhabitants' commuting patterns have not yet been chosen and utility owners are only now beginning to plan for the power plants that will be needed to light up those communities. Today's notoriously inefficient energy system can be replaced if the world gives unprecedented attention to energy efficiency. Dramatic changes are plausible over the next 50 years because so much of the energy canvas is still blank.

To make the task of reducing emissions vivid, we sliced the stabilization triangle into seven equal pieces, or "wedges," each representing one billion tons a year of averted emissions 50 years from now (starting from zero today). For example, a car driven 10,000 miles a year with a fuel efficiency of 30 miles per gallon (mpg) emits close to one ton of carbon annually. Transport experts predict that two billion cars will be zipping along the world's roads in 2056, each driven an average of 10,000

Managing the Climate Problem

At the present rate of growth, emissions of carbon dioxide will double by 2056 (below left). Even if the world then takes action to level them off, the atmospheric concentration of the gas will be headed above 560 parts per million, double the preindustrial value (below right)—a level widely regarded as capable of triggering severe climate changes. But if the world flattens out emissions beginning now and later ramps them down, it should be able to keep concentration substantially below 560 ppm.

Annual Emissions
In between the two emissions paths is the "stabilization triangle." It represents the total emissions cut that climate-friendly technologies must achieve in the coming 50 years.

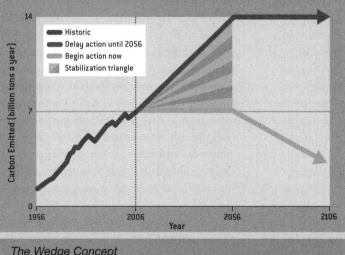

- Historic
- Delay action until 2056
- Begin action now
- Stabilization triangle

Carbon Emitted (billion tons a year)

14

7

0

1956   2006   2056   2106

Year

Cumulative Amount
Each part per million of CO_2 corresponds to a total of 2.1 billion tons of atmospheric carbon. Therefore, the 560-ppm level would mean about 1,200 billion tons, up from the current 800 billion tons. The difference of 400 billion tons actually allows for roughly 800 billion tons of emissions, because half the CO_2 emitted into the atmosphere enters the planet's oceans and forests. The two concentration trajectories shown here match the two emissions paths at the left.

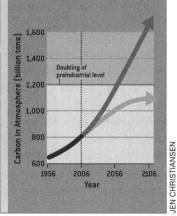

Carbon in Atmosphere (billion tons)

1,600

1,400

Doubling of preindustrial level

1,200

1,000

800

600

1956   2006   2056   2106

Year

The Wedge Concept
The stabilization triangle can be divided into seven "wedges," each a reduction of 25 billion tons of carbon emissions over 50 years. The wedge has proved to be a useful unit because its size and time frame match what specific technologies can achieve. Many combinations of technologies can fill the seven wedges.

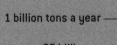

1 billion tons a year

25 billion tons total

50 years

miles a year. If their average fuel efficiency were 30 mpg, their tailpipes would spew two billion tons of carbon that year. At 60 mpg, they would give off a billion tons. The latter scenario would therefore yield one wedge.

Wedges

In our framework, you are allowed to count as wedges only those differences in two 2056 worlds that result from deliberate carbon policy. The current pace of emissions growth already includes some steady reduction in carbon intensity. The goal is to reduce it even more. For instance, those who believe that cars will average 60 mpg in 2056 even in a world that pays no attention to carbon cannot count this improvement as a wedge, because it is already implicit in the baseline projection.

Moreover, you are allowed to count only strategies that involve the scaling up of technologies already commercialized somewhere in the world. You are not allowed to count pie in the sky. Our goal in developing the wedge framework was to be pragmatic and realistic—to propose engineering our way out of the problem and not waiting for the cavalry to come over the hill. We argued that even with these two counting rules, the world can fill all seven wedges—and in several different ways [see sidebar, "15 Ways to Make a Wedge," page 56]. Individual countries—operating within a framework of international cooperation—will decide which wedges to pursue, depending on their institutional and economic capacities, natural resource endowments and political predilections.

To be sure, achieving nearly every one of the wedges requires new science and engineering to squeeze down costs and address the problems that inevitably accompany widespread deployment of new technologies. But holding CO_2 emissions in 2056 to their present rate, without choking off economic growth, is a desirable outcome within our grasp.

Ending the era of conventional coal-fired power plants is at the very top of the decarbonization agenda. Coal has become more competitive as a source of power and fuel because of energy security concerns and because of an increase in the cost of oil and gas. That is a problem because a coal power plant burns twice as much carbon per unit of electricity as a natural gas plant. In the absence of a concern about carbon, the world's coal utilities could build a few thousand large (1,000-megawatt) conventional coal plants in the next 50 years. Seven hundred such plants emit one wedge's worth of carbon. Therefore, the world could take some big steps toward the target of freezing emissions by not building those plants. The time to start is now. Facilities built in this decade could easily be around in 2056.

Efficiency in electricity use is the most obvious substitute for coal. Of the 14 billion tons of carbon emissions projected for 2056, perhaps six billion will come from producing power, mostly from coal. Residential and commercial buildings account for 60 percent of global electricity demand today (70 percent in the U.S.) and will consume most of the new power. So cutting buildings' electricity use in half—by equipping them with superefficient lighting and appliances—could lead to two wedges. Another wedge would be achieved if industry finds additional ways to use electricity more efficiently.

Decarbonizing the Supply

Even after energy-efficient technology has penetrated deeply, the world will still need power plants. They can be coal plants, but they will need to be carbon-smart ones that capture the CO_2 and pump it into the ground [see "Can We Bury Global Warming?" page 44]. Today's high oil prices are lowering the cost of the transition to this technology, because captured CO_2 can often be sold to an oil company that injects it into oil fields to squeeze out more oil; thus, the higher the price of oil, the

15 Ways to Make a Wedge

An overall carbon strategy for the next half a century produces seven wedges' worth of emissions reductions. Here are 15 technologies from which those seven can be chosen (taking care to avoid double-counting). Each of these measures, when phased in over 50 years, prevents the release of 25 billion tons of carbon. Leaving one wedge blank symbolizes that this list is by no means exhaustive.

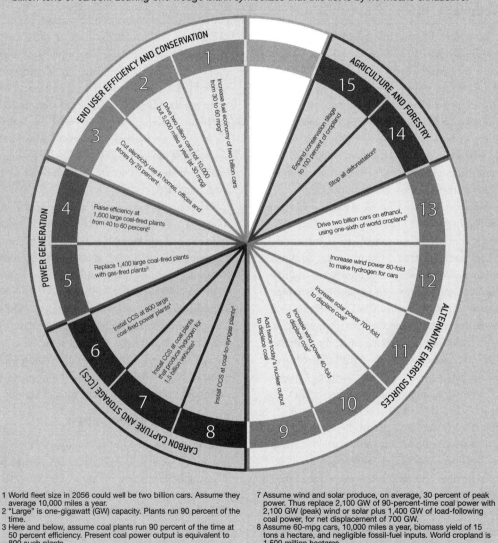

END USER EFFICIENCY AND CONSERVATION

1 Increase fuel economy of two billion cars from 30 to 60 mpg[1]

2 Drive two billion cars not 10,000 but 5,000 miles a year (at 30 mpg)[1]

3 Cut electricity use in homes, offices and stores by 25 percent

POWER GENERATION

4 Raise efficiency at 1,600 large coal-fired plants from 40 to 60 percent[2]

5 Replace 1,400 large coal-fired plants with gas-fired plants[3]

CARBON CAPTURE AND STORAGE (CCS)

6 Install CCS at 800 large coal-fired power plants[4]

7 Install CCS at coal plants that produce hydrogen for 1.5 billion vehicles[5]

8 Install CCS at coal-to-syngas plants[6]

9 Add twice today's nuclear output to displace coal

10 Increase wind power 40-fold to displace coal[7]

11 Increase solar power 700-fold to displace coal[7]

12 Increase wind power 80-fold to make hydrogen for cars

ALTERNATIVE ENERGY SOURCES

13 Drive two billion cars on ethanol, using one-sixth of world cropland[8]

AGRICULTURE AND FORESTRY

14 Stop all deforestation[9]

15 Expand conservation tillage to 100 percent of cropland

1 World fleet size in 2056 could well be two billion cars. Assume they average 10,000 miles a year.

2 "Large" is one-gigawatt (GW) capacity. Plants run 90 percent of the time.

3 Here and below, assume coal plants run 90 percent of the time at 50 percent efficiency. Present coal power output is equivalent to 800 such plants.

4 Assume 90 percent of CO_2 is captured.

5 Assume a car (10,000 miles a year, 60 miles per gallon equivalent) requires 170 kilograms of hydrogen a year.

6 Assume 30 million barrels of synfuels a day, about a third of today's total oil production. Assume half of carbon originally in the coal is captured.

7 Assume wind and solar produce, on average, 30 percent of peak power. Thus replace 2,100 GW of 90-percent-time coal power with 2,100 GW (peak) wind or solar plus 1,400 GW of load-following coal power, for net displacement of 700 GW.

8 Assume 60-mpg cars, 10,000 miles a year, biomass yield of 15 tons a hectare, and negligible fossil-fuel inputs. World cropland is 1,500 million hectares.

9 Carbon emissions from deforestation are currently about two billion tons a year. Assume that by 2056 the rate falls by half in the business-as-usual projection and to zero in the flat path.

JANET CHAO

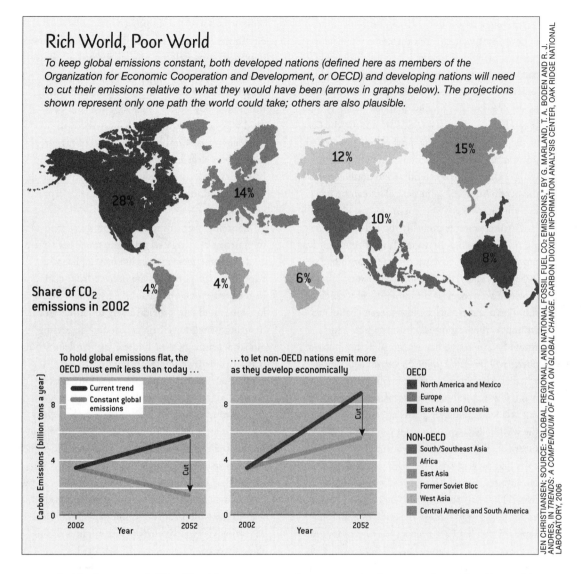

Rich World, Poor World

To keep global emissions constant, both developed nations (defined here as members of the Organization for Economic Cooperation and Development, or OECD) and developing nations will need to cut their emissions relative to what they would have been (arrows in graphs below). The projections shown represent only one path the world could take; others are also plausible.

28%

14%

12%

15%

10%

8%

4%

4%

6%

Share of CO$_2$ emissions in 2002

To hold global emissions flat, the OECD must emit less than today ...

Carbon Emissions (billion tons a year)

Current trend
Constant global emissions

Cut

2002   Year   2052

...to let non-OECD nations emit more as they develop economically

Cut

2002   Year   2052

OECD
North America and Mexico
Europe
East Asia and Oceania

NON-OECD
South/Southeast Asia
Africa
East Asia
Former Soviet Bloc
West Asia
Central America and South America

JEN CHRISTIANSEN; SOURCE: "GLOBAL, REGIONAL, AND NATIONAL FOSSIL FUEL CO$_2$ EMISSIONS," BY G. MARLAND, T. A. BODEN AND R. J. ANDRES, IN *TRENDS: A COMPENDIUM OF DATA ON GLOBAL CHANGE.* CARBON DIOXIDE INFORMATION ANALYSIS CENTER, OAK RIDGE NATIONAL LABORATORY, 2006

more valuable the captured CO$_2$. To achieve one wedge, utilities need to equip 800 large coal plants to capture and store nearly all the CO$_2$ otherwise emitted. Even in a carbon-constrained world, coal mining and coal power can stay in business, thanks to carbon capture and storage.

The large natural gas power plants operating in 2056 could capture and store their CO$_2$, too, per-

haps accounting for yet another wedge. Renewable and nuclear energy can contribute as well. Renewable power can be produced from sunlight directly, either to energize photovoltaic cells or, using focusing mirrors, to heat a fluid and drive a turbine. Or the route can be indirect, harnessing hydropower and wind power, both of which rely on sun-driven weather patterns. The intermittency of renewable power does not diminish its capacity to contribute wedges; even if coal and natural gas plants provide the backup power, they run only part-time (in tandem with energy storage) and use less carbon than if they ran all year. Not strictly renewable, but also usually included in the family, is geothermal energy, obtained by mining the heat in the earth's interior. Any of these sources, scaled up from its current contribution, could produce a wedge. One must be careful not to double-count the possibilities; the same coal plant can be left unbuilt only once.

Nuclear power is probably the most controversial of all the wedge strategies. If the fleet of nuclear power plants were to expand by a factor of five by 2056, displacing conventional coal plants, it would provide two wedges. If the current fleet were to be shut down and replaced with modern coal plants without carbon capture and storage, the result would be *minus* one-half wedge. Whether nuclear power will be scaled up or down will depend on whether governments can find political solutions to waste disposal and on whether plants can run without accidents. (Nuclear plants are mutual hostages: the world's least well-run plant can imperil the future of all the others.) Also critical will be strict rules that prevent civilian nuclear technology from becoming a stimulus for nuclear weapons development. These rules will have to be uniform across all countries, so as to remove the sense of a double standard that has long been a spur to clandestine facilities.

Oil accounted for 43 percent of global carbon emissions from fossil fuels in 2002, while coal accounted for 37 percent; natural gas made up the remainder. More than half the oil was used for transport. So smartening up electricity production alone cannot fill the stabilization triangle; transportation, too, must be decarbonized. As with coal-fired electricity, at least a wedge may be available from each of three complementary options: reduced use, improved efficiency and decarbonized energy sources. People can take fewer unwanted trips (telecommuting instead of vehicle commuting) and pursue the travel they cherish (adventure, family visits) in fuel-efficient vehicles running on low-carbon fuel. The fuel can be a product of crop residues or dedicated crops, hydrogen made from low-carbon electricity, or low-carbon electricity itself, charging an onboard battery. Sources of the low-carbon electricity could include wind, nuclear power, or coal with capture and storage.

Looming over this task is the prospect that, in the interest of energy security, the transport system could become *more* carbon intensive. That will happen if transport fuels are derived from coal instead of petroleum. Coal-based synthetic fuels, known as synfuels, provide a way to reduce global demand for oil, lowering its cost and decreasing global dependence on Middle East petroleum. But it is a decidedly climate-unfriendly strategy. A synfuel-powered car emits the same amount of CO_2 as a gasoline-powered car, but synfuel fabrication from coal spews out far more carbon than does refining gasoline from crude oil—enough to double the emissions per mile of driving. From the perspective of mitigating climate change, it is fortunate that the emissions at a synfuels plant can be captured and stored. If business-as-usual trends did lead to the widespread adoption of synfuel, then capturing CO_2 at synfuels plants might well produce a wedge.

Not all wedges involve new energy technology. If all the farmers in the world practiced no-till agriculture rather than conventional plowing, they would contribute a wedge. Eliminating deforestation would result in two wedges, if the alternative were for deforestation to continue at current rates. Curtailing emissions of methane, which today contribute about half as much to greenhouse warming as CO_2, may provide more than one wedge:

needed is a deeper understanding of the anaerobic biological emissions from cattle, rice paddies and irrigated land. Lower birth rates can produce a wedge, too—for example, if they hold the global population in 2056 near eight billion people when it otherwise would have grown to nine billion.

Action Plan

What set of policies will yield seven wedges? To be sure, the dramatic changes we anticipate in the fossil fuel system, including routine use of CO_2 capture and storage, will require institutions that reliably communicate a price for present and future carbon emissions. We estimate that the price needed to jump-start this transition is in the ballpark of $100 to $200 per ton of carbon—the range that would make it cheaper for owners of coal plants to capture and store CO_2 rather than vent it. The price might fall as technologies climb the learning curve. A carbon emissions price of $100 per ton is comparable to the current U.S. production credit for new renewable and nuclear energy relative to coal, and it is about half the current U.S. subsidy of ethanol relative to gasoline. It also was the price of CO_2 emissions in the European Union's emissions trading system for nearly a year, spanning 2005 and 2006. (One ton of carbon is carried in 3.7 tons of carbon dioxide, so this price is also $27 per ton of CO_2.) Based on carbon content, $100 per ton of carbon is $12 per barrel of oil and $60 per ton of coal. It is 25 cents per gallon of gasoline and two cents per kilowatt-hour of electricity from coal.

But a price on CO_2 emissions, on its own, may not be enough. Governments may need to stimulate the commercialization of low-carbon technologies to increase the number of competitive options available in the future. Examples include wind, photovoltaic power and hybrid cars. Also appropriate are policies designed to prevent the construction of long-lived capital facilities that are mismatched to future policy. Utilities, for instance, need to be encouraged to invest in CO_2 capture and storage for new coal power plants, which would be very costly to retrofit later. Still another set of policies can harness the capacity of energy producers to promote efficiency—motivating power utilities to care about the installation and maintenance of efficient appliances, natural gas companies to care about the buildings where their gas is burned and oil companies to care about the engines that run on their fuel.

To freeze emissions at the current level, if one category of emissions goes up, another must come down. If emissions from natural gas increase, the combined emissions from oil and coal must decrease. If emissions from air travel climb, those from some other economic sector must fall. And if today's poor countries are to emit more, today's richer countries must emit less.

How much less? It is easy to bracket the answer. Currently, the industrial nations—the members of the Organization for Economic Cooperation and Development (OECD)—account for almost exactly half the planet's CO_2 emissions, and the developing countries plus the nations formerly part of the Soviet Union account for the other half. In a world of constant total carbon emissions, keeping the OECD's share at 50 percent seems impossible to justify in the face of the enormous pent-up demand for energy in the non-OECD countries, where more than 80 percent of the world's people live. On the other hand, the OECD member states must emit *some* carbon in 2056. Simple arithmetic indicates that to hold global emissions rates steady, non-OECD emissions cannot even double.

One intermediate value results if all OECD countries were to meet the emissions-reduction target for the UK that was articulated in 2003 by Prime Minister Tony Blair—namely, a 60 percent reduction by 2050, relative to recent levels. The

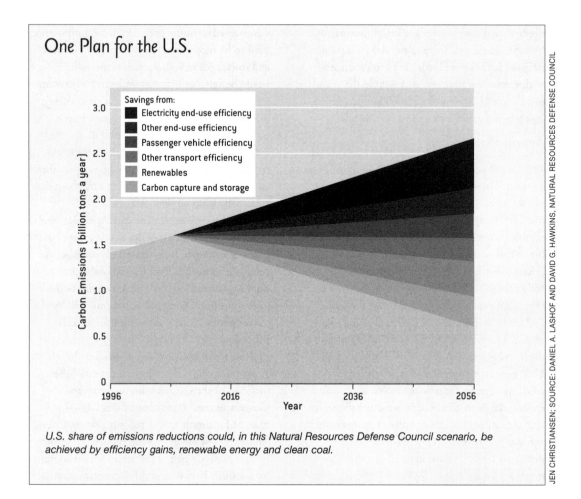

One Plan for the U.S.

Savings from:
- Electricity end-use efficiency
- Other end-use efficiency
- Passenger vehicle efficiency
- Other transport efficiency
- Renewables
- Carbon capture and storage

Carbon Emissions (billion tons a year)

Year

U.S. share of emissions reductions could, in this Natural Resources Defense Council scenario, be achieved by efficiency gains, renewable energy and clean coal.

JEN CHRISTIANSEN; SOURCE: DANIEL A. LASHOF AND DAVID G. HAWKINS, NATURAL RESOURCES DEFENSE COUNCIL

non-OECD countries could then emit 60 percent more CO_2. On average, by midcentury they would have one half the per capita emissions of the OECD countries. The CO_2 output of every country, rich or poor today, would be well below what it is generally projected to be in the absence of climate policy. In the case of the U.S., it would be about four times less.

Blair's goal would leave the average American emitting twice as much as the world average, as opposed to five times as much today. The U.S. could meet this goal in many ways [see illustration, "One Plan for the U.S.," above]. These strategies will be followed by most other countries as well. The resultant cross-pollination will lower every country's costs.

Fortunately, the goal of decarbonization does not conflict with the goal of eliminating the world's most extreme poverty. The extra carbon emissions produced when the world's nations accelerate the delivery of electricity and modern cooking fuel to the earth's poorest people can be compensated for by, at most, one fifth of a wedge of emissions reductions elsewhere.

Beyond 2056

The stabilization triangle deals only with the first 50-year leg of the future. One can imagine a relay race made of 50-year segments, in which the first runner passes a baton to the second in 2056. Intergenerational equity requires that the two runners have roughly equally difficult tasks. It seems to us that the task we have given the second runner (to cut the 2056 emissions rate in half between 2056 and 2106) will not be harder than the task of the first runner (to keep global emissions in 2056 at present levels)—provided that between now and 2056 the world invests in research and development to get ready. A vigorous effort can prepare the revolutionary technologies that will give the second half of the century a running start. Those options could include scrubbing CO_2 directly from the air, carbon storage in minerals, nuclear fusion, nuclear thermal hydrogen, and artificial photosynthesis. Conceivably, one or more of these technologies may arrive in time to help the first runner, although, as we have argued, the world should not count on it.

As we look back from 2056, if global emissions of CO_2 are indeed no larger than today's, what will have been accomplished? The world will have confronted energy production and energy efficiency at the consumer level, in all economic sectors and in economies at all levels of development. Buildings and lights and refrigerators, cars and trucks and planes, will be transformed. Transformed, also, will be the ways we use them.

The world will have a fossil fuel energy system about as large as today's but one that is infused with modern controls and advanced materials and that is almost unrecognizably cleaner. There will be integrated production of power, fuels and heat; greatly reduced air and water pollution; and extensive carbon capture and storage. Alongside the fossil energy system will be a nonfossil energy system approximately as large. Extensive direct and indirect harvesting of renewable energy will have brought about the revitalization of rural areas and the reclamation of degraded lands. If nuclear power is playing a large role, strong international enforcement mechanisms will have come into being to control the spread of nuclear technology from energy to weapons. Economic growth will have been maintained; the poor and the rich will both be richer. And our descendants will not be forced to exhaust so much treasure, innovation and energy to ward off rising sea level, heat, hurricanes and drought.

Critically, a planetary consciousness will have grown. Humanity will have learned to address its collective destiny—and to share the planet.

—SEPTEMBER 2006

What to Do About Coal

Cheap, plentiful coal is expected to fuel power plants for the foreseeable future, but can we keep it from devastating the environment?

DAVID G. HAWKINS, DANIEL A. LASHOF
AND ROBERT H. WILLIAMS

More than most people realize, dealing with climate change means addressing the problems posed by emissions from coal-fired power plants. Unless humanity takes prompt action to strictly limit the amount of carbon dioxide (CO_2) released into the atmosphere when consuming coal to make electricity, we have little chance of gaining control over global warming.

Coal—the fuel that powered the Industrial Revolution—is a particularly worrisome source of energy, in part because burning it produces consid-erably more carbon dioxide per unit of electricity generated than burning either oil or natural gas does. In addition, coal is cheap and will remain abundant long after oil and natural gas have become very scarce. With coal plentiful and inexpensive, its use is burgeoning in the U.S. and elsewhere and is expected to continue rising in areas with abundant coal resources. Indeed, U.S. power providers are expected to build the equivalent of nearly two hundred eighty 500-megawatt, coal-fired electricity plants between 2003 and 2030. Meanwhile, China is already constructing the equivalent of one large coal-fueled power station a week. Over their roughly 60-year life spans, the new generating facilities in operation by 2030 could collectively introduce into the atmosphere about as much carbon dioxide as was released by all the coal burned since the dawn of the Industrial Revolution.

Coal's projected popularity is disturbing not only for those concerned about climate change but also for those worried about other aspects of the environment and about human health and safety. Coal's market price may be low, but the true costs of its extraction, processing and consumption are high. Coal use can lead to a range of harmful con-sequences, including decapitated mountains, air

Overview

- Coal is widely burned for power but produces large quantities of climate-changing carbon dioxide.

- Compared with conventional power plants, new gasification facilities can more effectively and affordably extract CO_2 so it can be safely stored underground.

- The world must begin implementing carbon capture and storage soon to stave off global warming.

pollution from acidic and toxic emissions and water fouled with coal wastes. Extraction also endangers and can kill miners. Together such effects make coal production and conversion to useful energy one of the most destructive activities on the planet [see sidebar, "Coal's Toll," below].

Coal's Toll

Despite the current popularity of the term "clean coal," coal is, in fact, dirty. Although carbon capture and storage could prevent much carbon dioxide from entering the atmosphere, coal production and consumption is still one of the most destructive industrial processes. As long as the world consumes coal, more must be done to mitigate the harm it causes.

MINING DANGERS

Coal mining is among the most dangerous occupations. Official reports for 2005 indicate that roughly 6,000 people died (16 a day) in China from coal mine floods, cave-ins, fires and explosions. Unofficial estimates are closer to 10,000. Some 600,000 Chinese coal miners suffer from black lung disease. The U.S. has better safety practices than China and achieved an all-time low of 22 domestic fatalities in 2005. U.S. mines are far from perfect, however, as evidenced by a series of fatalities in early 2006.

ENVIRONMENTAL EFFECTS

Conventional coal mining, processing and transportation practices scar the landscape and pollute the water, which harms people and ecosystems. The most destructive mining techniques clear forests and blast away mountaintops. The "overburden" removed when a coal seam is uncovered is typically dumped into nearby valleys, where it often buries rivers and streams. Strip-mining operations rip apart ecosystems and reshape the landscape. Although regulations require land reclamation in principle, it is often left incomplete. As forests are replaced with nonnative grasslands, soils become compacted and streams contaminated. Underground mining can cause serious problems on the surface. Mines collapse and cause land subsidence, damaging homes and roads. Acidic mine drainage caused by sulfur compounds leaching from coal waste into surface waters has tainted thousands of streams. The acid leachate releases heavy metals that foul groundwater.

TOXIC EMISSIONS

Coal-fired power plants account for more than two-thirds of sulfur dioxide and about one-fifth of nitrogen oxide emissions in the U.S. Sulfur dioxide reacts in the atmosphere to form sulfate particles, which, in addition to causing acid rain, contribute to fine particulate pollution, a contaminant linked to thousands of premature deaths from lung disease nationwide. Nitrogen oxides combine with hydrocarbons to form smog-causing ground-level ozone. Coal-burning plants also emit approximately 48 metric tons of mercury a year in America. This highly toxic element persists in the ecosystem. After transforming into methyl mercury, it accumulates in the tissues of fishes. Ingested mercury is particularly detrimental to fetuses and young infants exposed during periods of rapid brain growth, causing developmental and neurological damage.

In keeping with *Scientific American*'s focus on climate concerns in this book, we will concentrate below on methods that can help prevent CO_2 generated during coal conversion from reaching the atmosphere. It goes without saying that the environmental, safety and health effects of coal production and use must be reduced as well. Fortunately, affordable techniques for addressing CO_2 emissions and these other problems already exist, although the will to implement them quickly still lags significantly.

Geologic Storage Strategy

The techniques that power providers could apply to keep most of the carbon dioxide they produce from entering the air are collectively called CO_2 capture and storage (CCS) or geologic carbon sequestration. These procedures involve separating out much of the CO_2 that is created when coal is converted to useful energy and transporting it to sites where it can be stored deep underground in porous media—mainly in depleted oil or gas fields or in saline formations (permeable geologic strata filled with salty water) [see "Can We Bury Global Warming?" page 44].

All the technological components needed for CCS at coal conversion plants are commercially ready—having been proved in applications unrelated to climate change mitigation, although integrated systems have not yet been constructed at the necessary scales. Capture technologies have been deployed extensively throughout the world both in the manufacture of chemicals (such as fertilizer) and in the purification of natural gas supplies contaminated with carbon dioxide and hydrogen sulfide ("sour gas"). Industry has gained considerable experience with CO_2 storage in operations that purify natural gas (mainly in Canada) as well as with CO_2 injection to boost oil production (primarily in the U.S.). Enhanced oil recovery processes account for most of the CO_2 that has been sent into underground reservoirs. Currently, about 35 million metric tons are injected annually to coax more petroleum out of mature fields, accounting for about 4 percent of U.S. crude oil output.

Implementing CCS at coal-consuming plants is imperative if the carbon dioxide concentration in the atmosphere is to be kept at an acceptable level. The 1992 United Nations Framework Convention on Climate Change calls for stabilizing the atmospheric CO_2 concentration at a "safe" level, but it does not specify what the maximum value should be. The current view of many scientists is that atmospheric CO_2 levels must be kept below 450 parts per million by volume (ppmv) to avoid unacceptable climate changes. Realization of this aggressive goal requires that the power industry start commercial-scale CCS projects within the next few years and expand them rapidly thereafter. This stabilization benchmark cannot be realized by CCS alone but can plausibly be achieved if it is combined with other eco-friendly measures, such as wide improvements in energy efficiency and much expanded use of renewable energy sources.

The Intergovernmental Panel on Climate Change (IPCC) estimated in 2005 that it is highly probable that geologic media worldwide are capable of sequestering at least two trillion metric tons of CO_2—more than is likely to be produced by fossil fuel–consuming plants during the 21st century. Society will want to be sure, however, that potential sequestration sites are evaluated carefully for their ability to retain CO_2 before they are allowed to operate. Two classes of risks are of concern: sudden escape and gradual leakage.

Rapid outflow of large amounts of CO_2 could be lethal to those in the vicinity. Dangerous sudden releases—such as that which occurred in 1986 at Lake Nyos in Cameroon, when CO_2 of volcanic origin asphyxiated 1,700 nearby villagers and thousands of cattle—are improbable for engineered CO_2 storage projects in carefully selected, deep porous geologic formations, according to the IPCC.

Gradual seepage of carbon dioxide into the air is also an issue, because over time it could defeat the goal of CCS. The 2005 IPCC report estimated that the fraction retained in appropriately selected and managed geologic reservoirs is very likely to exceed 99 percent over 100 years and likely to exceed 99 percent over 1,000 years. What remains to be demonstrated is whether in practice operators can routinely keep CO_2 leaks to levels that avoid unacceptable environmental and public health risks.

Technology Choices

Design studies indicate that existing power generation technologies could capture from 85 to 95 percent of the carbon in coal as CO_2, with the rest released to the atmosphere.

The coal conversion technologies that come to dominate will be those that can meet the objectives of climate change mitigation at the least cost. Fundamentally different approaches to CCS would be pursued for power plants using the conventional pulverized-coal steam cycle and the newer integrated gasification combined cycle (IGCC). Although today's coal IGCC power (with CO_2 venting) is slightly more expensive than coal steam-electric power, it looks like IGCC is the most effective and least expensive option for CCS.

Standard plants burn coal in a boiler at atmospheric pressure. The heat generated in coal combustion transforms water into steam, which turns a steam turbine, whose mechanical energy is converted to electricity by a generator. In modern plants the gases produced by combustion (flue gases) then pass through devices that remove particulates and oxides of sulfur and nitrogen before being exhausted via smokestacks into the air.

Carbon dioxide could be extracted from the flue gases of such steam-electric plants after the removal of conventional pollutants. Because the flue gases contain substantial amounts of nitrogen (the result of burning coal in air, which is about 80 percent nitrogen), the carbon dioxide would be recovered at low concentration and pressure—which implies that the CO_2 would have to be removed from large volumes of gas using processes that are both energy intensive and expensive. The captured CO_2 would then be compressed and piped to an appropriate storage site.

In an IGCC system coal is not burned but rather partially oxidized (reacted with limited quantities of oxygen from an air separation plant, and with steam) at high pressure in a gasifier. The product of gasification is so-called synthesis gas, or syngas, which is composed mostly of carbon monoxide and hydrogen, undiluted with nitrogen. In current practice, IGCC operations remove most conventional pollutants from the syngas and then burn it to turn both gas and steam turbine-generators in what is called a combined cycle.

In an IGCC plant designed to capture CO_2, the syngas exiting the gasifier, after being cooled and cleaned of particles, would be reacted with steam to produce a gaseous mixture made up mainly of carbon dioxide and hydrogen. The CO_2 would then be extracted, dried, compressed and transported to a storage site. The remaining hydrogen-rich gas would be burned in a combined-cycle plant to generate power [see sidebar, "Extracting and Storing Carbon Dioxide," page 66].

Analyses indicate that carbon dioxide capture at IGCC plants consuming high-quality bituminous coals would entail significantly smaller energy and cost penalties and lower total generation costs than what could be achieved in conventional coal plants that captured and stored CO_2. Gasification systems recover CO_2 from a gaseous stream at high concentration and pressure, a feature that makes the process much easier than it would be in conventional steam facilities. (The extent of these is less clear for lower-grade sub-bituminous coals and lignites, which have received much less study.)

Extracting and Storing Carbon Dioxide

To slow climate change, the authors urge power providers to build integrated gasification combined cycle (IGCC) coal power plants with carbon dioxide capture and storage (CCS) capabilities (below) rather than conventional steam-electric facilities. Conventional coal plants burn the fuel to transform water into steam to turn a turbine-generator. If CCS technology were applied to a steam plant, CO_2 would be extracted from the flue exhaust. An IGCC plant, in contrast, employs a partial oxidation reaction using limited oxygen to convert the coal into a so-called synthesis gas, or syngas (mostly hydrogen and carbon monoxide). It is much easier and less costly to remove CO_2 from syngas than from the flue gases of a steam plant. The hydrogen-rich syngas remaining after CO_2 extraction is then burned to run both gas and steam turbine-generators. The world's first commercial IGCC project that will sequester CO_2 underground is being planned near Long Beach, Calif.

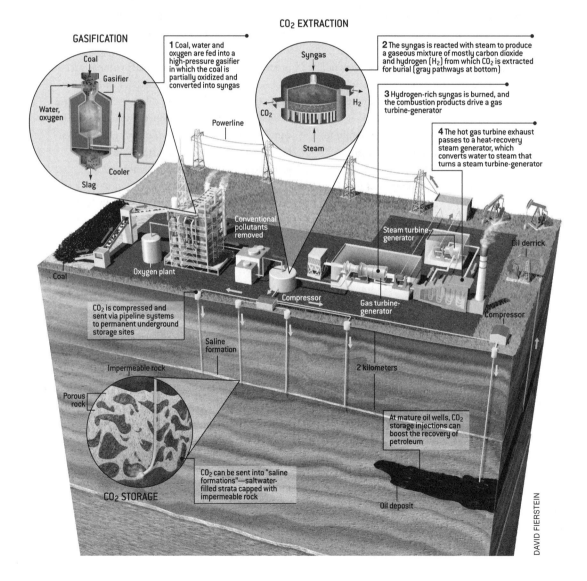

CO₂ EXTRACTION

GASIFICATION

Coal
Gasifier
Water, oxygen
Cooler
Slag

1 Coal, water and oxygen are fed into a high-pressure gasifier in which the coal is partially oxidized and converted into syngas

Syngas
CO_2
H_2
Steam

2 The syngas is reacted with steam to produce a gaseous mixture of mostly carbon dioxide and hydrogen (H_2) from which CO_2 is extracted for burial (gray pathways at bottom)

3 Hydrogen-rich syngas is burned, and the combustion products drive a gas turbine-generator

4 The hot gas turbine exhaust passes to a heat-recovery steam generator, which converts water to steam that turns a steam turbine-generator

Powerline
Conventional pollutants removed
Steam turbine-generator
Oil derrick
Coal
Oxygen plant
Compressor
Gas turbine-generator
Compressor

CO_2 is compressed and sent via pipeline systems to permanent underground storage sites

Saline formation
2 kilometers

Impermeable rock
Porous rock

At mature oil wells, CO_2 storage injections can boost the recovery of petroleum

CO_2 can be sent into "saline formations"—saltwater-filled strata capped with impermeable rock

CO₂ STORAGE

Oil deposit

DAVID FIERSTEIN

Precombustion removal of conventional pollutants, including mercury, makes it feasible to realize very low levels of emissions at much reduced costs and with much smaller energy penalties than with cleanup systems for flue gases in conventional plants.

Captured carbon dioxide can be transported by pipeline up to several hundred kilometers to suitable geologic storage sites and subsequent subterranean storage with the pressure produced during capture. Longer distances may, however, require recompression to compensate for friction losses during pipeline transfer.

Overall, pursuing CCS for coal power facilities requires the consumption of more coal to generate a kilowatt-hour of electricity than when CO_2 is vented—about 30 percent extra in the case of coal steam-electric plants and less than 20 percent more for IGCC plants. But overall coal use would not necessarily increase, because the higher price of coal-based electricity resulting from adding CCS equipment would dampen demand for coal-based electricity, making renewable energy sources and energy-efficient products more desirable to consumers.

The cost of CCS will depend on the type of power plant, the distance to the storage site, the properties of the storage reservoir and the availability of opportunities (such as enhanced oil recovery) for selling the captured CO_2. A recent study coauthored by one of us (Williams) estimated the incremental electric generation costs of two alternative CCS options for coal IGCC plants under typical production, transport and storage conditions. For CO_2 sequestration in a saline formation 100 kilometers from a power plant, the study calculated that the incremental cost of CCS would be 1.9 cents per kilowatt-hour (beyond the generation cost of 4.7 cents per kilowatt-hour for a coal IGCC plant that vents CO_2—a 40 percent premium). For CCS pursued in conjunction with enhanced oil recovery at a distance of 100 kilometers from the conversion plant, the analysis finds no increase in net generation cost would occur as long as the oil price is at least $35 per barrel, which is much lower than current prices.

CCS Now or Later?

Many electricity producers in the industrial world recognize that environmental concerns will at some point force them to implement CCS if they are to continue to employ coal. But rather than building plants that actually capture and store carbon dioxide, most plan to construct conventional steam facilities they claim will be "CO_2 capture ready"—convertible when CCS is mandated.

Power providers often defend those decisions by noting that the U.S. and most other countries with coal-intensive energy economies have not yet instituted policies for climate change mitigation that would make CCS cost-effective for uses not associated with enhanced oil recovery. Absent revenues from sales to oil field operators, applying CCS to new coal plants using current technology would be the least-cost path only if the cost of emitting CO_2 were at least $25 to $30 per metric ton. Many current policy proposals for climate change mitigation in the U.S. envision significantly lower cost penalties to power providers for releasing CO_2 (or similarly, payments for CO_2 emissions-reduction credits).

Yet delaying CCS at coal power plants until economy-wide carbon dioxide control costs are greater than CCS costs is shortsighted. For several reasons, the coal and power industries and society would ultimately benefit if deployment of plants fitted with CCS equipment were begun now.

First, the fastest way to reduce CCS costs is via "learning by doing"—the accumulation of experience in building and running such plants. The faster the understanding is accumulated, the quicker the know-how with the new technology will grow, and the more rapidly the costs will drop.

Second, installing CCS equipment as soon as possible should save money in the long run. Most power stations currently under construction will still be operating decades from now, when it is likely that CCS efforts will be obligatory. Retrofitting generating facilities for CCS is inherently more expensive than deploying CCS in new plants. Moreover, in the absence of CO_2 emission limits, familiar conventional coal steam-electric technologies will tend to be favored for most new plant construction over newer gasification technologies, for which CCS is more cost-effective.

Finally, rapid implementation would allow for continued use of fossil fuels in the near term (until more environmentally friendly sources become prevalent) without pushing atmospheric carbon dioxide beyond tolerable levels. Our studies indicate that it is feasible to stabilize atmospheric CO_2 levels at 450 ppmv over the next half a century if coal-based energy is completely decarbonized and other measures are implemented. This effort would

SEE   *Figure 6 in color section.*

involve decarbonizing 36 gigawatts of new coal-generating capacity by 2020 (corresponding to 7 percent of the new coal capacity expected to be built worldwide during the decade beginning in 2011 under business-as-usual conditions). In the 35 years after 2020, CO_2 capture would need to rise at an average rate of about 12 percent a year. Such a sustained pace is high compared with typical market growth rates for energy but is not unprecedented. It is much less than the expansion rate for nuclear generating capacity in its heyday—1956 to 1980—during which global capacity rose at an average rate of 40 percent annually. Further, the expansion rates for both wind and solar photovoltaic power capacities worldwide have hovered around 30 percent a year since the early 1990s. In

all three cases, such growth would not have been practical without public policy measures to support them.

Our calculations indicate that the costs of CCS deployment would be manageable as well. Using conservative assumptions—such as that technology will not improve over time—we estimate that the present worth of the cost of capturing and storing all CO_2 produced by coal-based electricity generation plants during the next 200 years will be $1.8 trillion (in 2002 dollars). That might seem like a high price tag, but it is equivalent to just 0.07 percent of the current value of gross world product over the same interval. Thus, it is plausible that a rapid decarbonization path for coal is both physically and economically feasible, although detailed regional analyses are needed to confirm this conclusion.

Policy Push Is Needed

Those good reasons for commencing concerted CCS efforts soon will probably not move the industry unless it is also prodded by new public policies. Such initiatives would be part of a broader drive to control carbon dioxide emissions from all sources.

In the U.S., a national program to limit CO_2 emissions must be enacted soon to introduce the government regulations and market incentives necessary to shift investment to the least polluting energy technologies promptly and on a wide scale. Leaders in the American business and policy communities increasingly agree that quantifiable and enforceable restrictions on global warming emissions are imperative and inevitable. To ensure that power companies put into practice the reductions in a cost-effective fashion, a market for trading CO_2 emissions credits should be created—one similar to that for the sulfur emissions that cause acid rain. In such a plan, organizations that intend to exceed designated emission limits may buy credits from others that are able to stay below these values.

Enhancing energy efficiency efforts and raising renewable energy production are critical to achieving carbon dioxide limits at the lowest possible cost. A portion of the emission allowances created by a carbon cap-and-trade program should be allocated to the establishment of a fund to help overcome institutional barriers and technical risks that obstruct widespread deployment of otherwise cost-effective CO_2 mitigation technologies.

Even if a carbon dioxide cap-and-trade program were enacted in the next few years the economic value of CO_2 emissions reduction may not be enough initially to convince power providers to invest in power systems with CCS. To avoid the construction of another generation of conventional coal plants, it is essential that the federal government establish incentives that promote CCS.

One approach would be to insist that an increasing share of total coal-based electricity generation comes from facilities that meet a low CO_2 emissions standard—perhaps a maximum of 30 grams of carbon per kilowatt-hour (an achievable goal using today's coal CCS technologies). Such a goal might be achieved by obliging electricity producers that use coal to include a growing fraction of decarbonized coal power in their supply portfolios. Each covered electricity producer could either generate the required amount of decarbonized coal power or purchase decarbonized-generation credits. This system would share the incremental costs of CCS for coal power among all U.S. coal-based electricity producers and consumers.

If the surge of conventional coal-fired power plants currently on drawing boards is built as planned, atmospheric carbon dioxide levels will almost certainly exceed 450 ppmv. We can meet global energy needs while still stabilizing CO_2 at 450 ppmv, however, through a combination of improved efficiency in energy use, greater reliance on renewable energy resources and, for the new coal investments that are made, the installation of CO_2 capture and geologic storage technologies. Even though there is no such thing as "clean coal," more can and must be done to reduce the dangers and environmental degradations associated with coal production and use. An integrated low-carbon energy strategy that incorporates CO_2 capture and storage can reconcile substantial use of coal in the coming decades with the imperative to prevent catastrophic changes to the earth's climate.

—SEPTEMBER 2006

Soccer Goes Green

At the World Cup, a new way to offset carbon emissions.

GUNJAN SINHA

Soccer, beer and bratwurst were very likely the only things on fans' minds as they descended on Germany to celebrate the World Cup this June. But all that partying had a downside—pollution. One million soccer tourists consumed a lot of energy. Environmentalism is part of the German zeitgeist, so it is only fitting that the event had a "green goal," too. A consortium including FIFA, the international soccer federation, and the German football association DFB donated 1.2 million euros to make this year's play-off the first sporting event to offset its carbon dioxide emissions by investing in three renewable energy projects.

Among the environmentally savvy, carbon-offset programs are the latest rage. At least a dozen companies offer the promise to mitigate greenhouse gas emissions from activities such as flying and driving and from events such as weddings and record releases. Voluntary offset programs, however, are not regulated, so consumers cannot be sure that they are investing in environmentally sound projects. What is more, reductions do not help much, because emissions from such activities are tiny: for instance, Germany's total carbon dioxide emissions are about 800 million tons per year—the World Cup emitted a mere 100,000 tons extra.

But governments are starting to pay heed to offsets. Europe established a cap-and-trade system last year that limits carbon dioxide emissions from about 50 percent of industry to reach its emissions goals as dictated by the Kyoto Protocol. Officials modeled the system on the sulfur dioxide trading market established in the U.S. in 1995, which has successfully cut levels of acid rain. As the trading market evolves, some environmentalists think that voluntary offset programs could join existing cap-and-trade market schemes to cut emissions even more substantially.

Right now, however, groups involved in voluntary projects are busy establishing credibility. "We wanted to serve as a model," says Christian Hochfeld of the Öko Institute in Berlin, an environmental think tank that developed the Green Goal project for the World Cup.

To that end, the institute chose projects that met criteria established by the World Wildlife Fund to better define high-quality development projects for industries affected by Kyoto. In Tamil Nadu, India, Women for Sustainable Development, a nonprofit organization, will oversee the installation of 700 to 1,000 biogas reactors—simple enclosed pits about the size of a well into which villagers dump cow dung. The fermenting mass generates gas, which fuels stoves and replaces kerosene. Two other sustainable energy projects will take place in South Africa. One will capture off-gas at a sewage treatment facility and burn it to supply electricity to Sebokeng, a township near Johannesburg. The other will replace a citrus farm's coal-fired heating system with one that burns sawdust—a by-product of wood processing

A Work in Progress

The European Union instituted caps on greenhouse gas emissions for about 50 percent of Europe's industries to meet Kyoto Protocol goals. That system suffered a setback in May after governments realized that they had set emissions limits too high. Most companies came in far below their limits, rendering their pollution credits worthless and eliminating financial incentive for them to cut carbon dioxide output. The European Union is discussing how to tighten the scheme.

The U.S. is not part of the Kyoto treaty, but states are picking up the slack. Under the Regional Greenhouse Gas Initiative, Northeast and mid-Atlantic states plan to implement a market-based cap-and-trade program for all greenhouse gases emitted from power plants in the area. In California, a pending bill would cap greenhouse gas emissions from all industries in the state.

usually discarded. The projects will offset all the soccer tournament's emissions.

World Cup organizers could have planted trees—as has been done by previous sporting events, such as the Super Bowl—or invested in other projects on home turf. But planting has been criticized because trees take years of growth to suck up an equivalent amount of released carbon. Also, not all renewable energy projects constitute an "offset." If an undertaking that would have happened anyway was jump-started through government subsidies—using wind power, for example—it cannot be considered a true offset, environmental groups say.

Currently, voluntary activities do not generate tradable emission credits. But imagine if they did. Suppose, for example, that anyone could earn credits for cutting consumption or increasing efficiency. Those villagers in India could earn credits for reducing their emissions that they could in turn sell, says Annie Petsonk, international counsel at Environmental Defense, a New York City–based nonprofit group. "How interesting would it be to have everyone participating? It would stimulate so much energy efficiency," she predicts. "We're talking about tapping economic power in favor of protecting the environment on a huge scale."

—AUGUST 2006

PART TWO

(THE FUTURE OF ENERGY)

NUCLEAR: SECOND TIME AROUND

Next-Generation Nuclear Power

New, safer and more economical nuclear reactors could not only satisfy many of our future energy needs but could combat global warming as well.

JAMES A. LAKE, RALPH G. BENNETT AND JOHN F. KOTEK

Rising electricity prices and last summer's rolling blackouts in California have focused fresh attention on nuclear power's key role in keeping America's lights on. Today 103 nuclear plants crank out a fifth of the nation's total electrical output. And despite residual public misgivings over Three Mile Island and Chernobyl, the industry has learned its lessons and established a solid safety record during the past decade. Meanwhile the efficiency and reliability of nuclear plants have climbed to record levels. Now with the ongoing debate about reducing greenhouse gases to avoid the potential onset of global warming, more people are recognizing that nuclear reactors produce electricity without discharging into the air carbon dioxide or pollutants such as nitrogen oxides and smog-causing sulfur compounds. The world demand for energy is projected to rise by about 50 percent by 2030 and to nearly double by 2050. Clearly, the time seems right to reconsider the future of nuclear power [see sidebar, "The Case for Nuclear Power," pages 86–88].

No new nuclear plant has been ordered in the U.S. since 1978, nor has a plant been finished since 1995. Resumption of large-scale nuclear plant construction requires that challenging questions be addressed regarding the achievement of economic viability, improved operating safety, efficient waste management and resource utilization, as well as weapons nonproliferation, all of which are influenced by the design of the nuclear reactor system that is chosen.

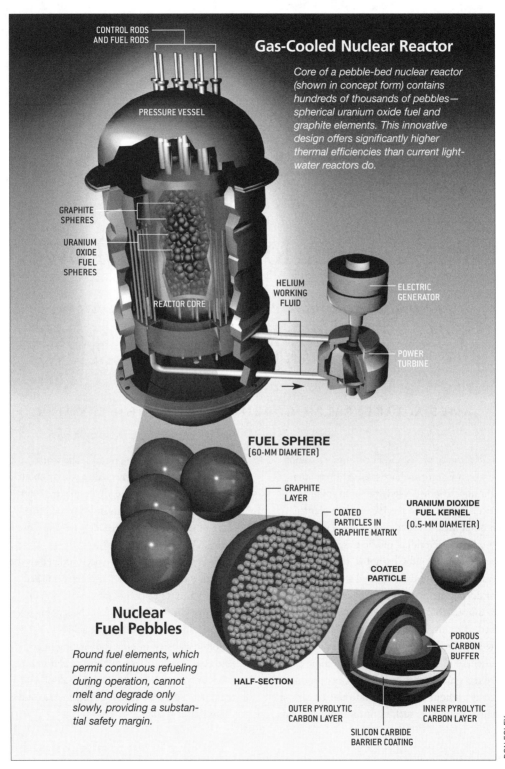

Gas-Cooled Nuclear Reactor

Core of a pebble-bed nuclear reactor (shown in concept form) contains hundreds of thousands of pebbles— spherical uranium oxide fuel and graphite elements. This innovative design offers significantly higher thermal efficiencies than current light-water reactors do.

CONTROL RODS AND FUEL RODS

PRESSURE VESSEL

GRAPHITE SPHERES

URANIUM OXIDE FUEL SPHERES

REACTOR CORE

HELIUM WORKING FLUID

ELECTRIC GENERATOR

POWER TURBINE

FUEL SPHERE
(60-MM DIAMETER)

GRAPHITE LAYER

COATED PARTICLES IN GRAPHITE MATRIX

URANIUM DIOXIDE FUEL KERNEL
(0.5-MM DIAMETER)

COATED PARTICLE

HALF-SECTION

POROUS CARBON BUFFER

OUTER PYROLYTIC CARBON LAYER

INNER PYROLYTIC CARBON LAYER

SILICON CARBIDE BARRIER COATING

Nuclear Fuel Pebbles

Round fuel elements, which permit continuous refueling during operation, cannot melt and degrade only slowly, providing a substantial safety margin.

DON FOLEY

Designers of new nuclear systems are adopting novel approaches in the attempt to attain success. First, they are embracing a system-wide view of the nuclear fuel cycle that encompasses all steps from the mining of ore through the management of wastes and the development of the infrastructure to support these steps. Second, they are evaluating systems in terms of their sustainability—meeting present needs without jeopardizing the ability of future generations to prosper. It is a strategy that helps to illuminate the relation between energy supplies and the needs of the environment and society. This emphasis on sustainability can lead to the development of nuclear energy–derived products besides electrical power, such as hydrogen fuel for transportation. It also promotes the exploration of alternative reactor designs and nuclear fuel–recycling processes that could yield significant reductions in waste while recovering more of the energy contained in uranium.

We believe that wide-scale deployment of nuclear power technology offers substantial advantages over other energy sources yet faces significant challenges regarding the best way to make it fit into the future.

Future Nuclear Systems

In response to the difficulties in achieving sustainability, a sufficiently high degree of safety and a competitive economic basis for nuclear power, the U.S. Department of Energy initiated the Generation IV program in 1999. Generation IV refers to the broad division of nuclear designs into four categories: early prototype reactors (Generation I), the large central station nuclear power plants of today (Generation II), the advanced light-water reactors and other systems with inherent safety features that have been designed in recent years (Generation III), and the next-generation systems to be designed and built two decades from now (Generation IV) [see sidebar, "Nuclear Power Primer,"

page 78]. By 2000 international interest in the Generation IV project had resulted in a nine-country coalition that includes Argentina, Brazil, Canada, France, Japan, South Africa, South Korea, the UK and the U.S. Participating states are mapping out and collaborating on the research and development of future nuclear energy systems.

Although the Generation IV program is exploring a wide variety of new systems, a few examples serve to illustrate the broad approaches reactor designers are developing to meet their objectives. These next-generation systems are based on three general classes of reactors: gas-cooled, water-cooled and fast-spectrum.

Gas-Cooled Reactors

Nuclear reactors using gas (usually helium or carbon dioxide) as a core coolant have been built and operated successfully but have achieved only limited use to date. An especially exciting prospect known as the pebble-bed modular reactor possesses many design features that go a good way toward meeting Generation IV goals. This gas-cooled system is being pursued by engineering teams in China, South Africa and the U.S. South Africa plans to build a full-size prototype and begin operation in 2006.

The pebble-bed reactor design is based on a fundamental fuel element, called a pebble, that is a billiard-ball-size graphite sphere containing about 15,000 uranium oxide particles with the diameter of poppy seeds [see illustrations, page 76]. The evenly dispersed particles each have several high-density coatings on them. One of the layers, composed of tough silicon carbide ceramic, serves as a pressure vessel to retain the products of nuclear fission during reactor operation or accidental temperature excursions. About 330,000 of these spherical fuel pebbles are placed into a metal vessel surrounded by a shield of graphite blocks. In addition, as many as 100,000 unfueled

Nuclear Power Primer

Most of the world's nuclear power plants are pressurized water reactors. In these systems, water placed under high pressure (155 atmospheres) to suppress boiling serves as both the coolant and the working fluid. Initially developed in the U.S. based on experience gained from the American naval reactor program, the first commercial pressurized light-water reactor commenced operation in 1957.

The reactor core of a pressurized water reactor is made up of arrays of zirconium alloy–clad fuel rods composed of small cylinders (pellets) of mildly enriched uranium oxide with the diameter of a dime. A typical 17-by-17-square array of fuel rods constitutes a fuel assembly, and about 200 fuel assemblies are arranged to form a reactor core. Cores, which are typically approximately 3.5 meters in diameter and 3.5 meters high, are contained within steel pressure vessels that are 15 to 20 centimeters thick.

The nuclear fission reactions produce heat that is removed by circulating water. The coolant is pumped into the core at about 290 degrees Celsius and exits the core at about 325 degrees C. To control the power level, control rods are inserted into the fuel arrays. Control rods are made of materials that moderate the fission reaction by absorbing the slow (thermal) neutrons emitted during fission. They are raised out of or lowered into the core to control the rate of the nuclear reaction. To change the fuel or in the case of an accident, the rods are lowered all the way into the core to shut down the reaction.

In the primary reactor coolant loop, the hot water exits the reactor core and flows through a heat exchanger (called a steam generator), where it gives up its heat to a secondary steam loop that operates at a lower pressure level. The steam produced in the heat exchanger is then expanded through a steam turbine, which in turn spins a generator to produce electricity (typically 900 to 1,100 megawatts). The steam is then condensed and pumped back into the heat exchanger to complete the loop. Aside from the source of heat, nuclear power plants are generally similar to coal or fuel-fired electrical generating facilities.

There are several variants of the light-water-cooled reactor, most notably boiling-water reactors, which operate at lower pressure (usually 70 atmospheres) and generate steam directly in the reactor core, thus eliminating the need for the intermediate heat exchanger. In a smaller number of nuclear power plants, the reactor coolant fluid is heavy water (containing the hydrogen isotope deuterium), carbon dioxide gas or a liquid metal such as sodium.

The reactor pressure vessel is commonly housed inside a concrete citadel that acts as a radiation shield. The citadel is in turn enclosed within a steel-reinforced concrete containment building. The containment building is designed to prevent leakage of radioactive gases or fluids in an accident.

graphite pebbles are loaded into the core to shape its power and temperature distribution by spacing out the hot fuel pebbles.

Heat-resistant refractory materials are used throughout the core to allow the pebble-bed system to operate much hotter than the 300 degrees Celsius temperatures typically produced in today's light-water-cooled (Generation II) designs. The helium working fluid, exiting the core at 900 degrees C, is fed directly into a gas turbine/generator system that generates electricity at a comparatively high 40 percent thermal efficiency level, one-quarter better than current light-water reactors.

The comparatively small size and the general simplicity of pebble-bed reactor designs add to their economic feasibility. Each power module, producing 120 megawatts of electrical output, can be deployed in a unit one-tenth the size of today's central station plants, which permits the development of more flexible, modest-scale projects that may offer more favorable economic results. For example, modular systems can be manufactured in the factory and then shipped to the construction site.

The pebble-bed system's relative simplicity compared with current designs is dramatic: these units have only about two dozen major plant subsystems, compared with about 200 in light-water reactors. Significantly, the operation of these plants can be extended into a temperature range that makes possible the low emissions production of hydrogen from water or other feedstocks for use in fuel cells and clean-burning transportation engines, technologies on which a sustainable hydrogen-based energy economy could be based [see sidebar, "The Case for Nuclear Power," page 86–88].

These next-generation reactors incorporate several important safety features as well. Being a noble gas, the helium coolant will not react with other materials, even at high temperatures. Further, because the fuel elements and reactor core are made of refractory materials, they cannot melt and will degrade only at the extremely high temperatures encountered in accidents (more than 1,600 degrees C), a characteristic that affords a considerable margin of operating safety.

Yet other safety benefits accrue from the continuous, on-line fashion in which the core is refueled: during operation, one pebble is removed from the bottom of the core about once a minute as a replacement is placed on top. In this way, all the pebbles gradually move down through the core like gumballs in a dispensing machine, taking about six months to do so. This feature means that the system contains the optimum amount of fuel for operation, with little extra fissile reactivity. It eliminates an entire class of excess-reactivity accidents that can occur in current water-cooled reactors. Also, the steady movement of pebbles through regions of high and low power production means that each experiences less extreme operating conditions on average than do fixed fuel configurations, again adding to the unit's safety margin. After use, the spent pebbles must be placed in long-term storage repositories, the same way that used-up fuel rods are handled today.

Water-Cooled Reactors

Even standard water-cooled nuclear reactor technology has a new look for the future. Aiming to overcome the possibility of accidents resulting from loss of coolant (which occurred at Three Mile Island) and to simplify the overall plant, a novel class of Generation IV systems has arisen in which all the primary components are contained in a single vessel. An American design in this class is the international reactor innovative and secure (IRIS) concept developed by Westinghouse Electric.

Housing the entire coolant system inside a damage-resistant pressure vessel means that the primary system cannot suffer a major loss of coolant even if one of its large pipes breaks. Because the pressure vessel will not allow fluids to escape, any resulting accident is limited to a much

Water-Cooled Nuclear Reactor

IRIS reactor design developed by Westinghouse Electric (depicted in conceptual form) is novel in that both the steam generator (heat exchanger) and the control rod actuator drives are enclosed within the thick steel pressure vessel.

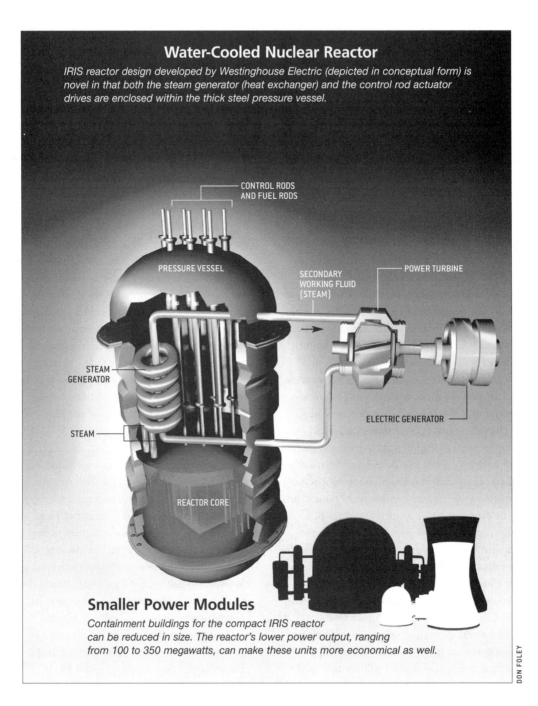

CONTROL RODS AND FUEL RODS

PRESSURE VESSEL

SECONDARY WORKING FLUID (STEAM)

POWER TURBINE

STEAM GENERATOR

ELECTRIC GENERATOR

STEAM

REACTOR CORE

Smaller Power Modules

Containment buildings for the compact IRIS reactor can be reduced in size. The reactor's lower power output, ranging from 100 to 350 megawatts, can make these units more economical as well.

DON FOLEY

Open and Closed Nuclear Fuel Cycles

"Once-through," or open, nuclear fuel cycle (shown shaded) takes uranium ore, processes it into fissile fuel, burns it a single time in a reactor and then disposes of it in a geological repository. This approach, which is employed the U.S., uses only 1 percent of the uranium's energy content. In a closed cycle (white line only), the spent fuel is processed to reclaim its uranium and plutonium fuel content for reuse. This recycling method is used today in France, Japan and the U.K. Future closed cycles based on fast-spectrum reactors could reclaim other actinides that are currently treated as waste.

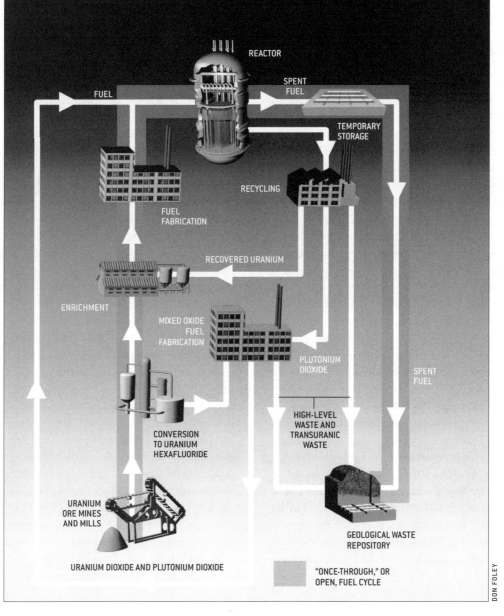

REACTOR

SPENT FUEL

FUEL

TEMPORARY STORAGE

RECYCLING

FUEL FABRICATION

RECOVERED URANIUM

ENRICHMENT

MIXED OXIDE FUEL FABRICATION

PLUTONIUM DIOXIDE

SPENT FUEL

CONVERSION TO URANIUM HEXAFLUORIDE

HIGH-LEVEL WASTE AND TRANSURANIC WASTE

URANIUM ORE MINES AND MILLS

GEOLOGICAL WASTE REPOSITORY

URANIUM DIOXIDE AND PLUTONIUM DIOXIDE

"ONCE-THROUGH," OR OPEN, FUEL CYCLE

DON FOLEY

more moderate drop in pressure than could occur in previous designs. To accomplish this compact configuration, several important simplifications are incorporated in these reactors. The subsystems within the vessel are stacked to enable passive heat transfer by natural circulation during accidents. In addition, the control rod drives are located in the vessel, eliminating the chance that they could be ejected from the core. These units can also be built as small power modules, thereby allowing more flexible and lower-cost deployment.

Designers of these reactors are also exploring the potential of operating plants at high temperature and pressure (more than 374 degrees C and 221 atmospheres), a condition known as the critical point of water, at which the distinction between liquid and vapor blurs. Beyond its critical point, water behaves as a continuous fluid with exceptional specific heat (thermal storage capacity) and superior heat transfer (thermal conductance) properties. It also does not boil as it heats up or flash to steam if it undergoes rapid depressurization. The primary advantage to operating above the critical point is that the system's thermal efficiency can reach as high as 45 percent and approach the elevated temperature regime at which hydrogen fuel production can become viable.

Although reactors based on supercritical water appear very similar to standard Generation II designs at first glance, the differences are many. For instance, the cores of the former are considerably smaller, which helps to economize on the pressure vessel and the surrounding plant. Next, the associated steam-cycle equipment is substantially simplified because it operates with a single-phase working fluid. In addition, the smaller core and the low coolant density reduce the volume of water that must be held within the containment vessel in the event of an accident. Because the low-density coolant does not moderate the energy of the neutrons, fast-spectrum reactor designs, with their asso-

ciated sustainability benefits, can be contemplated. The chief downside to supercritical water systems is that the coolant becomes increasingly corrosive. This means that new materials and methods to control corrosion and erosion must be developed. Supercritical water reactor research is ongoing in Canada, France, Japan, South Korea and the U.S.

Fast-Spectrum Reactors

A design approach for the longer term is the fast-spectrum (or high-energy neutron) reactor, another type of Generation IV system. An example of this class of reactor is being pursued by design teams in France, Japan, Russia, South Korea and elsewhere. The American fast-reactor development program was canceled in 1995, but U.S. interest might be revived under the Generation IV initiative.

Most nuclear reactors employ a thermal, or relatively low energy, neutron-emissions spectrum. In a thermal reactor the fast (high-energy) neutrons generated in the fission reaction are slowed down to "thermal" energy levels as they collide with the hydrogen in water or other light nuclides. Although these reactors are economical for generating electricity, they are not very effective in producing nuclear fuel (in breeder reactors) or recycling it.

Most fast-spectrum reactors built to date have used liquid sodium as the coolant. Future versions of this reactor class may utilize sodium, lead, a lead-bismuth alloy or inert gases such as helium or carbon dioxide. The higher-energy neutrons in a fast reactor can be used to make new fuel or to destroy long-lived wastes from thermal reactors and plutonium from dismantled weapons. By recycling the fuel from fast reactors, they can deliver much more energy from uranium while reducing the amount of waste that must be disposed of for the long term. These breeder-reactor designs are one of the keys to increasing the sustainability of future nuclear energy systems, especially if the use of nuclear energy is to grow significantly.

Fast-Spectrum Nuclear Reactor

Cores of fast-spectrum nuclear reactors such as General Electric's Super PRISM design (shown in conceptual form), which produce fast (high-energy) neutrons, are often cooled with molten metals. In breeder-reactor configurations, these high-energy neutrons are used to create nuclear fuel.

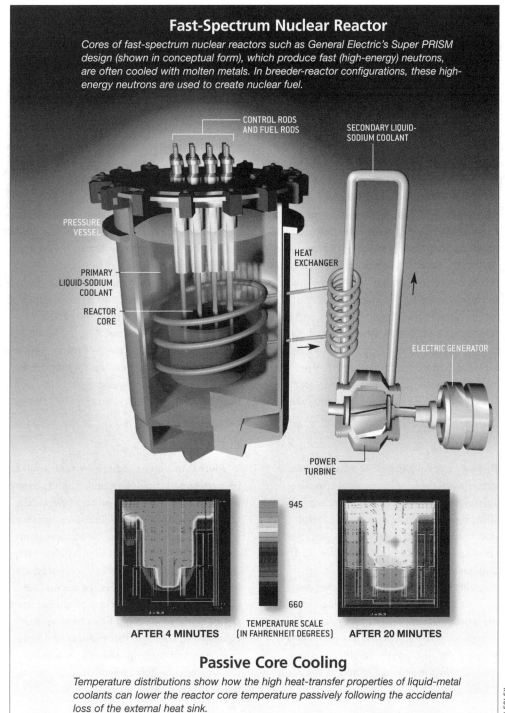

CONTROL RODS
AND FUEL RODS

SECONDARY LIQUID-
SODIUM COOLANT

PRESSURE
VESSEL

HEAT
EXCHANGER

PRIMARY
LIQUID-SODIUM
COOLANT

REACTOR
CORE

ELECTRIC GENERATOR

POWER
TURBINE

945

660

AFTER 4 MINUTES

TEMPERATURE SCALE
(IN FAHRENHEIT DEGREES)

AFTER 20 MINUTES

Passive Core Cooling

Temperature distributions show how the high heat-transfer properties of liquid-metal coolants can lower the reactor core temperature passively following the accidental loss of the external heat sink.

How Secure Are Nuclear Plants from Terrorists?

The tragic events of September 11, 2001, raise troubling questions about the vulnerability of nuclear facilities to terrorist attacks. Although stringent civilian and military security countermeasures have been implemented to stop determined assaults, the deliberate crash of a large commercial airliner looms in the imagination. So should Americans be worried? The answer is no and yes.

A nuclear power plant is not an easy target for an airliner flying at high speed, because an off-center hit on a domed, cylindrical containment building would not substantially affect the building structure. Located at or below grade, the reactor core itself is typically less than 10 feet in diameter and 12 feet high. It is enclosed in a heavy steel vessel surrounded by a concrete citadel. Reactor containment designs differ in their details, but in all cases they are meant to survive the worst of nature's forces (including earthquakes, tornadoes and hurri-

canes). Despite not being designed to resist acts of war, containment enclosures can withstand crashes of small aircraft.

Even though the reactor core is protected, some of the piping and reactor cooling equipment, the auxiliary apparatus and the adjacent switchyard may be vulnerable to a direct hit. Nuclear power stations, however, are outfitted with multiple emergency cooling systems, as well as with emergency power supplies, should power be disabled. In the improbable event that all of these backup precautions were destroyed, the reactor core could overheat and melt. But even in this extreme case, which is similar to what occurred at Three Mile Island, the radioactive core materials would still be contained within the pressure vessel.

If nuclear plants have an Achilles' heel, it is the on-site temporary storage facilities for spent nuclear fuel. Although these depositories usually contain several used fuel assemblies and there-

Beyond supporting the use of a fast-neutron spectrum, metal coolants have several attractive qualities. First, they possess exceptional heat-transfer properties, which allows metal-cooled reactors to withstand accidents like the ones that happened at Three Mile Island and Chernobyl. Second, some (but not all) liquid metals are considerably less corrosive to components than water is, thereby extending the operating life of reactor vessels and other critical subsystems. Third, these high-temperature systems can operate near atmospheric pressure, greatly simplifying system design and reducing potential industrial hazards in the plant.

More than a dozen sodium-cooled reactors have been operated around the world. This experience has called attention to two principal difficulties that must be overcome. Sodium reacts with water to generate high heat, a possible accident source. This characteristic has led sodium-cooled reactor designers to include a secondary sodium system to isolate the primary coolant in the reactor core from the water in the electricity-producing steam system. Some new designs focus on novel heat-exchanger technologies that guard against leaks.

The second challenge concerns economics. Because sodium-cooled reactors require two heat-transfer steps between the core and the turbine,

fore more total radioactivity than a reactor does, most of the more dangerous radioactive isotopes in the old fuel have already decayed away. This is particularly true for the gaseous fission products that could get into the air, whose half-lives can be measured in months. Spent fuel assemblies that have been removed relatively recently from reactors are kept in deep pools of water to cool them and shield the radiation they emit. These open-air pools are surrounded by thick-walled, steel-lined concrete containers. After a few years, the materials are transferred into concrete, air-cooled dry fuel-storage casks.

Although cooling pools provide a relatively small and, hence, difficult target for terrorists, a pinpoint attack could drain a pool's water, causing the fuel to overheat and melt. Experts say that a standard fire hose would be enough to refill the pool. Even if the fuel were to melt, little radioactive particulate would be produced that might become airborne, specialists say. An air-

liner crash into dry fuel-storage casks would probably just knock them aside. If any casks cracked, broken bits of oxidized fuel cladding could carry some radioactivity skyward, according to nuclear safety experts.

Some experts believe that the Nuclear Regulatory Commission will soon order the reinforcement of auxiliary nuclear plant equipment and waste storage facilities.

Should such a terrorist onslaught occur, plans are in place to evacuate nearby residents, although it must be said that critics claim these schemes to be impractical. It is thought, however, that there would be eight to 10 hours available to get out safely, long before evacuees received a significant radioactive dose. The most severe potential adverse effect could be long-term contamination of the local area by airborne particulates, which would be expensive to clean up.

—THE EDITORS

capital costs are increased and thermal efficiencies are lower than those of the most advanced gas and water-cooled concepts (about 38 percent in an advanced sodium-cooled reactor compared with 45 percent in a supercritical water reactor). Moreover, liquid metals are opaque, making inspection and maintenance of components more difficult.

Next-generation fast-spectrum reactor designs attempt to capitalize on the advantages of earlier configurations while addressing their shortcomings. The technology has advanced to the point at which it is possible to envision fast-spectrum reactors that engineers believe will pose little chance of a meltdown. Further, nonreactive coolants such as

inert gases, lead or lead-bismuth alloys may eliminate the need for a secondary coolant system and improve the approach's economic viability.

Nuclear energy has arrived at a crucial stage in its development. The economic success of the current generation of plants in the U.S. has been based on improved management techniques and careful practices, leading to growing interest in the purchase of new plants. Novel reactor designs can dramatically improve the safety, sustainability and economics of nuclear energy systems in the long term, opening the way to their widespread deployment.

—JANUARY 2002

The Case for Nuclear Power

Today 438 nuclear power plants generate about 16 percent of the world's electricity. In the U.S., 103 nuclear power plants provide about 20 percent of the country's electrical production. Although no new nuclear facilities have been ordered in the U.S. for more than two decades, the electrical output of U.S. generators has grown by almost 8 percent a year as the industry matured and became more efficient. In the past 10 years alone, American nuclear plants have added more than 23,000 megawatts—the equivalent of 23 large power plants—to the total electricity supply despite the lack of any new construction. In the meantime, the production increase has lowered the unit cost of nuclear power generation. This improvement has led to growing interest among the business community in extending plant operating licenses and perhaps purchasing new nuclear facilities.

It may be surprising to some that the use of nuclear energy has direct benefits to the environment, specifically air quality. Although debate continues about the potential for the disruption of the earth's climate by emissions of carbon dioxide and other greenhouse gases, there is no doubt about the serious health consequences of air pollution from the burning of fossil fuels. Unlike fossil fuel power plants, nuclear plants do not produce carbon dioxide, sulfur or nitrogen oxides. Nuclear power production in the U.S. annually avoids the emission of more than 175 million tons of carbon that would have been released into the environment if the same amount of electricity had instead been generated by burning coal.

Little attention has been paid to nuclear energy's capacity for producing hydrogen for use in transportation fuel cells and other cleaner power plants. A very straightforward approach is to use the energy from a high-temperature nuclear reactor to drive a steam-reforming reaction of methane. This process still creates carbon dioxide as a by-product, however. Several direct thermochemical reactions can give rise to hydrogen using water and high temperature. Research on the thermochemical decomposition of sulfuric acid and other hydrogen-forming reactions is under way in Japan and the U.S. The economics of nuclear-based hydrogen remain to be proved, but enormous potential exists for this route, perhaps operating in a new electricity-hydrogen cogeneration mode.

IMPROVING ECONOMICS

Any nuclear construction in the U.S. must address challenging economic issues concerning their capital costs and financing. The problem is that the current generation of nuclear power plants, represented by three Nuclear Regulatory Commission–certified advanced light-water reactor designs, costs about $1,500 per kilowatt electric (kWe) of generating capacity, which may not be sufficiently competitive to restart nuclear construction. A widely discussed cost goal for new (Generation III and IV) nuclear plant projects is $1,000 per kWe. Achievement of this aim would make them competitive (on a unit-cost basis) with the most economical alternative, the combined-cycle natural gas plant. Any next-generation facilities must in addition be completed within about three years to keep financing costs to a manageable level. New streamlined, but as yet untried, licensing procedures should speed the process.

Given the past experience with nuclear projects in the U.S., it will be difficult for designers and builders to meet these goals. To achieve the cost objective, nuclear engineers are seeking to attain higher thermal efficiencies by raising operating temperatures and simplifying subsystems and components. Speeding plant construction will require the standardization of plant designs, factory fabrication and certification procedures; the division of plants into smaller modules that avoid the need for on-site construction; and the use of computerized assembly-management techniques. In this way, the building work can be verified in virtual reality before it proceeds in the field.

ADVANCING SAFETY

As the economic performance of the nuclear power industry has improved over the past 20 years, so too has its safety performance. The Three Mile Island accident in 1979 focused the attention of plant owners and operators on the need to boost safety margins and performance. The number of so-called safety-significant events reported to the Nuclear Regulatory Commission, for example, averaged about two per plant per year in 1990 but had dropped to less than one tenth of that by 2000. In the meantime, public confidence in the safety of nuclear power has been largely restored since the Chernobyl accident in 1986, according to recent polls.

Long-term safety goals for next-generation nuclear facilities were formulated during the past year by international and domestic experts at the request of the U.S. Department of Energy. They established three major objectives: to improve the safety and reliability of plants, to lessen the possibility of significant damage during accidents and to minimize the potential consequences of any accidents that do occur. Accomplishing these aims will require new plant designs that incorporate inherent safety features to prevent accidents and to keep accidents from deteriorating into more severe situations that could release radioactivity into the environment.

NUCLEAR WASTE DISPOSAL AND REUSE

Outstanding issues regarding the handling and disposal of nuclear waste and safeguarding against nuclear proliferation must also be addressed. The Yucca Mountain long-term underground repository in Nevada is being evaluated to decide whether it can successfully accept spent commercial fuel. It is, however, a decade behind schedule and even when completed will not accommodate the quantities of waste projected for the future.

The current "once-through," or open, nuclear fuel cycle uses freshly mined uranium, burns it a single time in a reactor and then discharges it as waste. This approach results in only about 1 percent of the energy content of the uranium being converted to electricity. It also produces large volumes of spent nuclear fuel that must be disposed of in a safe fashion. Both these drawbacks can be avoided by recycling the spent fuel—that is, recovering the useful materials from it.

Most other countries with large nuclear power programs—including France, Japan and the UK—employ what is called a closed nuclear fuel cycle. In these countries, used fuel is recycled to recover uranium and plutonium (produced during irradiation in reactors) and reprocess it into new fuel. This effort doubles the amount of energy recovered from the fuel and removes most of the long-lived radioactive elements from the waste that must be permanently stored. It should be noted, though, that recycled fuel is today more expensive than newly mined fuel. Current recycling technology also leads to the separation of plutonium, which could potentially be diverted into weapons.

Essentially all nuclear fuel recycling is performed using a process known as PUREX (plutonium uranium extraction), which was initially developed for extracting pure plutonium for nuclear weapons. In PUREX recycling, used fuel assemblies are transported to a recycling plant in heavily shielded, damage-resistant shipping casks. The fuel assemblies are chopped up and dissolved by strong acids. The fuel solution then undergoes a solvent-extraction procedure to separate the fission products and other elements from the uranium and the plutonium, which are purified. The uranium and plutonium are used to fabricate mixed oxide fuel for use in light-water reactors.

Recycling helps to minimize the production of nuclear waste. To reduce the demand for storage space, a sustainable nuclear fuel cycle would separate the short-lived, high-heat-producing fission products, particularly cesium 137 and strontium 90. These elements would be held separately in convectively cooled facilities for 300 to 500 years, until they had decayed to safe levels. An optimized closed (fast-reactor) fuel cycle would recycle not just the uranium and plutonium but all actinides in the fuel, including neptunium,

americium and curium. In a once-through fuel cycle, more than 98 percent of the expected long-term radiotoxicity is caused by the resulting neptunium 237 and plutonium 242 (with half-lives of 2.14 million and 387,000 years, respectively). Controlling the long-term effects of a repository becomes simpler if these long-lived actinides are also separated from the waste and recycled. The removal of cesium, strontium and the actinides from the waste shipped to a geological repository could increase its capacity by a factor of 50.

Because of continuing interest in advancing the sustainability and economics of nuclear fuel cycles, several countries are developing more effective recycling technologies. Today an electrometallurgical process that precludes the separation of pure plutonium is under development in the U.S. at Argonne National Laboratory. Advanced aqueous recycling procedures that offer similar advantages are being studied in France, Japan and elsewhere.

ENSURING NONPROLIFERATION

A critical aspect of new nuclear energy systems is ensuring that they do not allow weapons-usable materials to be diverted from the reprocessing cycle. When nations acquire nuclear weapons, they usually develop dedicated facilities to produce fissile materials rather than collecting nuclear materials from civilian power plants. Commercial nuclear fuel cycles are generally the most costly and difficult route for production of weapons-grade materials. New fuel cycles must continue to be designed to guard against proliferation.

The Workings of an Ancient Nuclear Reactor

Two billion years ago parts of an African uranium deposit spontaneously underwent nuclear fission. The examples recovered shed light not only on possible changes in fundamental physical constants but also on how buried nuclear waste migrates over time.

ALEX P. MESHIK

Overview: Fossil Reactors

- Three decades ago French scientists discovered that parts of a uranium deposit then being mined in Gabon had long ago functioned as natural fission reactors.

- The author and two colleagues recently used measurements of xenon gas (a product of uranium fission) to deduce that one of these ancient reactors must have operated with a duty cycle of about half an hour on and at least two and a half hours off.

- Perhaps further studies of xenon retained within minerals will reveal natural nuclear reactors elsewhere. But for the moment, the examples discovered in Gabon remain unique windows on possible changes in fundamental physical constants and on how buried nuclear waste migrates over time.

In May 1972 a worker at a nuclear plant in France noticed something suspicious. He had been conducting a routine analysis of uranium derived from a seemingly ordinary source of ore. As is the case with all natural uranium, the material under study contained three isotopes—that is to say, three forms with differing atomic masses: uranium 238, the most abundant variety; uranium 234, the rarest; and uranium 235, the isotope that is coveted because it can sustain a nuclear chain reaction. Elsewhere in the earth's crust, on the moon and even in meteorites, uranium 235 atoms make up 0.720 percent of the total. But in these samples, which came from the Oklo deposit in Gabon (a former French colony in west equatorial Africa), uranium 235 constituted just 0.717 percent. That tiny discrepancy was enough to alert French scientists that something strange had happened. Further analyses showed that ore from at least one part of the mine was far short on uranium 235: some 200 kilograms appeared to be missing—enough to make half a dozen or so nuclear bombs.

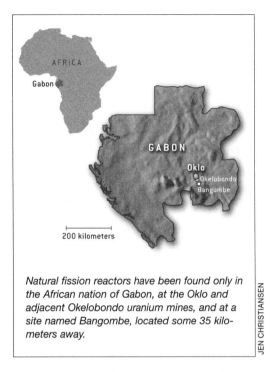

Natural fission reactors have been found only in the African nation of Gabon, at the Oklo and adjacent Okelobondo uranium mines, and at a site named Bangombe, located some 35 kilometers away.

JEN CHRISTIANSEN

For weeks, specialists at the French Atomic Energy Commission (CEA) remained perplexed. The answer came only when someone recalled a prediction published 19 years earlier. In 1953 George W. Wetherill of the University of California at Los Angeles and Mark G. Inghram of the University of Chicago pointed out that some uranium deposits might have once operated as natural versions of the nuclear fission reactors that were then becoming popular. Shortly thereafter, Paul K. Kuroda, a chemist from the University of Arkansas, calculated what it would take for a uranium ore body spontaneously to undergo self-sustained fission. In this process, a stray neutron causes a uranium 235 nucleus to split, which gives off more neutrons, causing others of these atoms to break apart in a nuclear chain reaction.

Kuroda's first condition was that the size of the uranium deposit should exceed the average length that fission-inducing neutrons travel, about two-thirds of a meter. This requirement helps to ensure that the neutrons given off by one fissioning nucleus are absorbed by another before escaping from the uranium vein.

A second prerequisite is that uranium 235 must be present in sufficient abundance. Today even the most massive and concentrated uranium deposit cannot become a nuclear reactor, because the uranium 235 concentration, at less than 1 percent, is just too low. But this isotope is radioactive and decays about six times faster than does uranium 238, which indicates that the fissile fraction was much higher in the distant past. For example, two billion years ago (about when the Oklo deposit formed) uranium 235 must have constituted approximately 3 percent, which is roughly the level provided artificially in the enriched uranium used to fuel most nuclear power stations.

The third important ingredient is a neutron "moderator," a substance that can slow the neutrons given off when a uranium nucleus splits so that they are more apt to induce other uranium nuclei to break apart. Finally, there should be no significant amounts of boron, lithium or other so-called poisons, which absorb neutrons and would thus bring any nuclear reaction to a swift halt.

Amazingly, the actual conditions that prevailed two billion years ago in what researchers eventually determined to be 16 separate areas within the Oklo and adjacent Okelobondo uranium mines were very close to what Kuroda outlined. These zones were all identified decades ago. But only recently did my colleagues and I finally clarify major details of what exactly went on inside one of those ancient reactors.

Proof in the Light Elements

Physicists confirmed the basic idea that natural fission reactions were responsible for the depletion in uranium 235 at Oklo quite soon after the anomalous uranium was discovered. Indisputable proof came from an examination of the new, lighter ele-

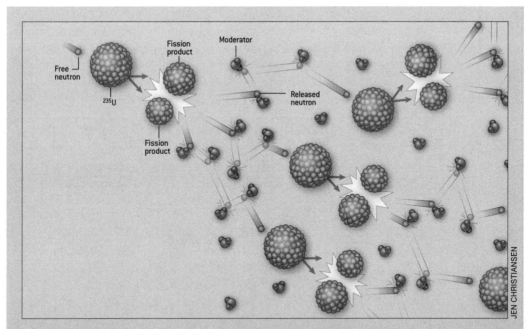

JEN CHRISTIANSEN

Fission Up Close

Nuclear chain reactions begin when a single free neutron strikes the nucleus of a fissile atom, such as uranium 235 (upper left). That nucleus then splits, creating two smaller atoms and releasing other neutrons, which fly off at great speed and must be slowed before they can induce other fissile nuclei to break apart. In the Oklo deposit, just as in today's light-water nuclear reactors, the slowing agent, or "moderator," was ordinary water. But the Oklo reactors differed from those built today in how they were regulated. Whereas nuclear power plants use neutron-absorbing control rods, the Oklo reactors just heated up until the water boiled away.

ments created when a heavy nucleus is broken in two. The abundance of these fission products proved so high that no other conclusion could be drawn. A nuclear chain reaction very much like the one that Enrico Fermi and his colleagues famously demonstrated in 1942 had certainly taken place, all on its own and some two billion years before.

Shortly after this astonishing discovery, physicists from around the world studied the evidence for these natural nuclear reactors and came together to share their work on "the Oklo phenomenon" at a special 1975 conference held in Libreville, the capital of Gabon. The next year George A. Cowan, who represented the U.S. at that meeting (and who, incidentally, is one of the founders of the renowned Santa Fe Institute, where he is still affiliated), wrote an article for *Scientific American* ["A Natural Fission Reactor," July 1976] in which he explained what scientists had surmised about the operation of these ancient reactors.

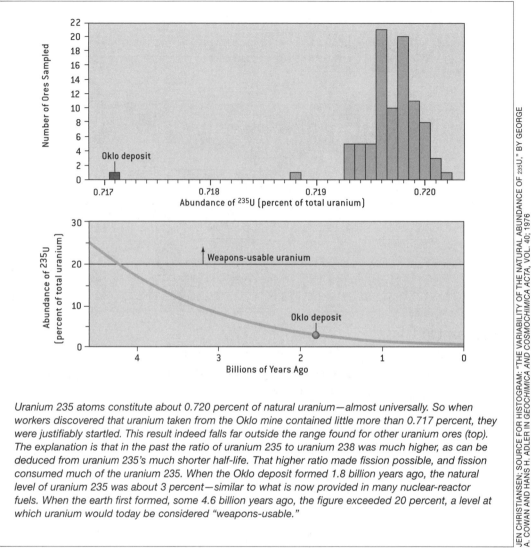

Uranium 235 atoms constitute about 0.720 percent of natural uranium—almost universally. So when workers discovered that uranium taken from the Oklo mine contained little more than 0.717 percent, they were justifiably startled. This result indeed falls far outside the range found for other uranium ores (top). The explanation is that in the past the ratio of uranium 235 to uranium 238 was much higher, as can be deduced from uranium 235's much shorter half-life. That higher ratio made fission possible, and fission consumed much of the uranium 235. When the Oklo deposit formed 1.8 billion years ago, the natural level of uranium 235 was about 3 percent—similar to what is now provided in many nuclear-reactor fuels. When the earth first formed, some 4.6 billion years ago, the figure exceeded 20 percent, a level at which uranium would today be considered "weapons-usable."

JEN CHRISTIANSEN; SOURCE FOR HISTOGRAM: "THE VARIABILITY OF THE NATURAL ABUNDANCE OF 235U," BY GEORGE A. COWAN AND HANS H. ADLER IN GEOCHIMICA AND COSMOCHIMICA ACTA, VOL. 40: 1976

Cowan described, for example, how some of the neutrons released during the fission of uranium 235 were captured by the more abundant uranium 238, which became uranium 239 and, after emitting two electrons, turned into plutonium 239. More than two tons of this plutonium isotope were generated within the Oklo deposit. Although almost all this material, which has a 24,000-year half-life, has since disappeared (primarily through natural radioactive decay), some of the plutonium itself underwent fission, as attested by the presence of its characteristic fission products. The abundance of those lighter elements allowed scientists to deduce that fission reactions must have gone on for hundreds of thousands of years. From the amount of uranium 235 consumed, they calculated the total energy released, 15,000 megawatt-years, and from this and other evidence were able to work out the average power output, which was probably less than 100 kilowatts—say, enough to run a few dozen toasters.

It is truly amazing that more than a dozen natural reactors spontaneously sprang into existence and that they managed to maintain a modest power output for perhaps a few hundred millennia. Why is it that these parts of the deposit did not explode and destroy themselves right after nuclear chain reactions began? What mechanism provided the necessary self-regulation? Did these reactors run steadily or in fits and starts? The solutions to these puzzles emerged slowly after initial discovery of the Oklo phenomenon. Indeed, the last question lingered for more than three decades before my colleagues and I at Washington University in St. Louis began to address it by examining a piece of this enigmatic African ore.

Noble-Gas Epiphanies

Our recent work on one of the Oklo reactors centered on an analysis of xenon, a heavy inert gas, which can remain imprisoned within minerals for billions of years. Xenon possesses nine stable isotopes, produced in various proportions by different nuclear processes. Being a noble gas, it resists chemical bonding with other elements and is thus easy to purify for isotopic analysis. Xenon is extremely rare, which allows scientists to use it to detect and trace nuclear reactions, even those that occurred in primitive meteorites before the solar system came into existence.

To analyze the isotopic composition of xenon requires a mass spectrometer, an instrument that can separate atoms according to their atomic weight. I was fortunate to have access to an extremely accurate xenon mass spectrometer, one built by my Washington colleague, Charles M. Hohenberg. But before using his apparatus, we had

to extract the xenon from our sample. Scientists usually just heat the host material, often above the melting point, so that the rock loses its crystalline structure and cannot hold onto its hidden cache of xenon. To glean greater information about the genesis and retention of this gas, we adopted a more delicate approach called laser extraction, which releases xenon selectively from a single mineral grain, leaving adjacent areas intact.

We applied this technique to many tiny spots on our lone available fragment of Oklo rock, only one millimeter thick and four millimeters across. Of course, we first needed to decide where exactly to aim the laser beam. Here Hohenberg and I relied on our colleague Olga Pravdivtseva, who had constructed a detailed X-ray map of our sample and identified the constituent minerals. After each extraction, we purified the resulting gas and passed the xenon into Hohenberg's mass spectrometer, which indicated the number of atoms of each isotope present.

Our first surprise was the location of the xenon. It was not, as we had expected, found to a significant extent in the uranium-rich mineral grains. Rather the lion's share was trapped in aluminum phosphate minerals, which contain no uranium at all. Remarkably, these grains showed the highest concentration of xenon ever found in any natural material. The second epiphany was that the extracted gas had a significantly different isotopic makeup from what is usually produced in nuclear reactors. It had seemingly lost a large portion of the xenon 136 and 134 that would certainly have been created from fission, whereas the lighter varieties of the element were modified to a lesser extent.

How could such a change in isotopic composition have come about? Chemical reactions would not do the trick, because all isotopes are chemically identical. Perhaps nuclear reactions, such as neutron capture? Careful analysis allowed my colleagues and me to reject this possibility as well. We

Xenon Reveals Cyclic Operation

Efforts to account for the isotopic composition of xenon at Oklo required a consideration of other elements as well. Iodine in particular drew attention because xenon arises from its radioactive decay. Modeling the creation of fission products and their radioactive decay revealed that the peculiar isotopic composition of xenon resulted from the cyclic operation of the reactor. That cycle is depicted in the three panels at the right.

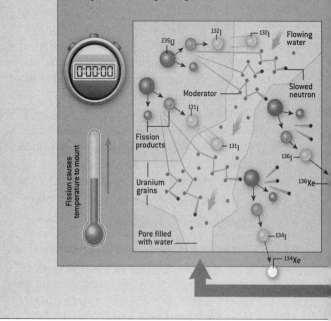

1 Groundwater permeating the deposit acted as a moderator, allowing the fission of uranium 235 to begin. The decay of certain radioactive fission products quickly gave rise to xenon 134 and 136, but these gas atoms had a tendency to be driven off by the rising heat of the reactor. The longer-lived xenon precursors iodine 131 and 132 went into solution and were washed away about as swiftly as they were created.

also considered the physical sorting of different isotopes that sometimes takes place: heavier atoms move a bit more slowly than their lighter counterparts and can thus sometimes separate from them. Uranium-enrichment plants—industrial facilities that require considerable skill to construct—take advantage of this property to produce reactor fuel. But even if nature could miraculously create a similar process on a microscopic scale, the mix of xenon isotopes in the aluminum phosphate grains we studied would have been different from what

we found. For example, measured with respect to the amount of xenon 132 present, the depletion of xenon 136 (being four atomic mass units heavier) would have been twice that of xenon 134 (two atomic mass units heavier) if physical sorting had operated. We did not see that pattern.

Our understanding of the anomalous composition of the xenon came only after we thought harder about how this gas was born. None of the xenon isotopes we measured were the direct result of uranium fission. Rather, they were the products of

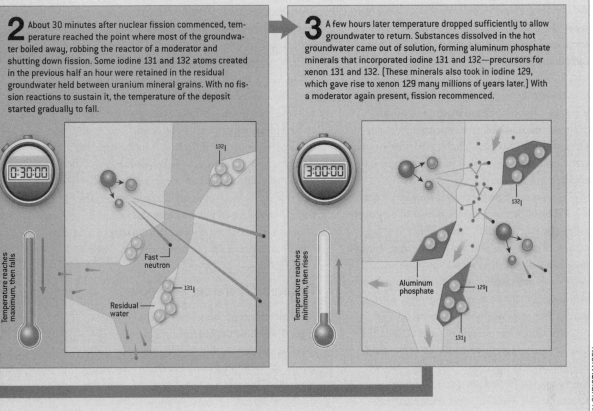

2 About 30 minutes after nuclear fission commenced, temperature reached the point where most of the groundwater boiled away, robbing the reactor of a moderator and shutting down fission. Some iodine 131 and 132 atoms created in the previous half an hour were retained in the residual groundwater held between uranium mineral grains. With no fission reactions to sustain it, the temperature of the deposit started gradually to fall.

3 A few hours later temperature dropped sufficiently to allow groundwater to return. Substances dissolved in the hot groundwater came out of solution, forming aluminum phosphate minerals that incorporated iodine 131 and 132—precursors for xenon 131 and 132. (These minerals also took in iodine 129, which gave rise to xenon 129 many millions of years later.) With a moderator again present, fission recommenced.

132 I

Temperature reaches maximum, then falls

Fast neutron

131 I

Residual water

Temperature reaches minimum, then rises

132 I

Aluminum phosphate

129 I

131 I

JEN CHRISTIANSEN

the decay of radioactive isotopes of iodine, which in turn were formed from radioactive tellurium and so forth, according to a well-known sequence of nuclear reactions that gives rise to stable xenon.

Our key insight was the realization that different xenon isotopes in our Oklo sample were created at different times—following a schedule that depended on the half-lives of their iodine parents and tellurium grandparents. The longer a particular radioactive precursor lives, the longer xenon formation from it is held off. For example, production of

xenon 136 began at Oklo only about a minute after the onset of self-sustained fission. An hour later the next lighter stable isotope, xenon 134, appeared. Then, some days after the start of fission, xenon 132 and 131 came on the scene. Finally, after millions of years, and well after the nuclear chain reactions terminated, xenon 129 formed.

Had the Oklo deposit remained a closed system, the xenon accumulated during operation of its natural reactors would have preserved the normal isotopic composition produced by fission. But scientists have no reason to think that the system was closed. Indeed, there is good cause to suspect the opposite. The evidence comes from a consideration of the simple fact that the Oklo reactors somehow regulated themselves. The most likely mechanism involves the action of groundwater, which presumably boiled away after the temperature reached some critical level. Without water present to act as a neutron moderator, nuclear chain reactions would have temporarily ceased. Only after things cooled off and sufficient groundwater once again permeated the zone of reaction could fission resume.

This picture of how the Oklo reactors probably worked highlights two important points: very likely they pulsed on and off in some fashion, and large quantities of water must have been moving through these rocks—enough to wash away some of the xenon precursors, tellurium and iodine, which are water soluble. The presence of water also helps to explain why most of the xenon now resides in grains of aluminum phosphate rather than in the uranium-rich minerals where fission first created these radioactive precursors. The xenon did not simply migrate from one set of preexisting minerals to another—it is unlikely that aluminum phosphate minerals were present before the Oklo reactors began operating. Instead those grains of aluminum phosphate probably formed in place through the action of the nuclear-heated water, once it had cooled to about 300 degrees Celsius.

During each active period of operation of an Oklo reactor and for some time afterward, while the temperature remained high, much of the xenon gas (including xenon 136 and 134, which were generated relatively quickly) was driven off. When the reactor cooled down, the longer-lived xenon precursors (those that would later spawn xenon 132, 131 and 129, which we found in relative abundance) were preferentially incorporated into growing grains of aluminum phosphate.

Then, as more water returned to the reaction zone, neutrons became properly moderated and fission once again resumed, allowing the cycle of heating and cooling to repeat. The result was the peculiar segregation of xenon isotopes we uncovered.

It is not entirely obvious what forces kept this xenon inside the aluminum phosphate minerals for almost half the planet's lifetime. In particular, why was the xenon generated during a given operational pulse not driven off during the next one? Presumably, it became imprisoned in the cagelike structure of the aluminum phosphate minerals, which were able to hold on to the xenon gas created within them, even at high temperatures. The details remain fuzzy, but whatever the final answers are, one thing is clear: the capacity of aluminum phosphate for capturing xenon is truly amazing.

Nature's Operating Schedule

After my colleagues and I had worked out in a general way how the observed set of xenon isotopes was created inside the aluminum phosphate grains, we attempted to model the process mathematically. This exercise revealed much about the timing of reactor operation, with all xenon isotopes providing pretty much the same answer. The Oklo reactor we studied had switched "on" for 30 minutes and "off" for at least 2.5 hours. The pattern is not unlike what one sees in some geysers, which slowly heat up, boil off their supply of groundwater in a spectacular display, refill, and repeat the cycle, day in and day out, year after year. This similarity supports the notion not only that groundwater passing through the Oklo deposit was a neutron moderator but also that its boiling away at times accounted for the self-regulation that protected these natural reactors from destruction. In this regard, it was extremely effective, allowing not a single meltdown or explosion during hundreds of thousands of years.

One would imagine that engineers working in the nuclear power industry could learn a thing or two from Oklo. And they certainly can, though not necessarily about reactor design. The more important lessons may be about how to handle nuclear waste. Oklo, after all, serves as a good analogue for a long-term geologic repository, which is why scientists have examined in great detail how the various products of fission have migrated away from these natural reactors over time. They have also scrutinized a similar zone of ancient nuclear fission found in exploratory boreholes drilled at a site called Bangombe, located some 35 kilometers away. The Bangombe reactor is of special interest because it was more shallowly buried than those unearthed at the Oklo and Okelobondo mines and thus has had more water moving through it in recent times. In all, the observations boost confidence that many kinds of dangerous nuclear waste can be successfully sequestered underground.

Oklo also demonstrates a way to store some forms of nuclear waste that were once thought to be almost impossible to prevent from contaminating the environment. Since the advent of nuclear power generation, huge amounts of radioactive xenon 135, krypton 85 and other inert gases that nuclear plants generate have been released into the atmosphere. Nature's fission reactors suggest the possibility of locking those waste products away in aluminum phosphate minerals, which have a unique ability to capture and retain such gases for billions of years.

The Oklo reactors may also teach scientists about possible shifts in what was formerly thought to be a fundamental physical constant, one called α (alpha), which controls such universal quantities as the speed of light [see "Inconstant Constants," by John D. Barrow and John K. Webb; *Scientific American*, June 2005]. For three decades, the two-billion-year-old Oklo phenomenon has been used to argue against α having changed. But last year Steven K. Lamoreaux and Justin R. Torgerson of Los Alamos National Laboratory drew on Oklo to posit that this "constant" has, in fact, varied significantly (and, strangely enough, in the opposite sense from what others have recently proposed). Lamoreaux and Torgerson's calculations hinge on certain details about how Oklo operated, and in that respect the work my colleagues and I have done might help elucidate this perplexing issue.

Were these ancient reactors in Gabon the only ones ever to have formed on the earth? Two billion years ago the conditions necessary for self-sustained fission must not have been too rare, so perhaps other natural reactors will one day be discovered. I expect that a few telltale wisps of xenon could aid immensely in this search.

—NOVEMBER 2005

Smarter Use of Nuclear Waste

*Fast-neutron reactors could extract much more energy
from recycled nuclear fuel, minimize the risks
of weapons proliferation and markedly reduce the time
nuclear waste must be isolated.*

WILLIAM H. HANNUM, GERALD E. MARSH
AND GEORGE S. STANFORD

Despite long-standing public concern about the safety of nuclear energy, more and more people are realizing that it may be the most environmentally friendly way to generate large amounts of electricity. Several nations, including Brazil, China, Egypt, Finland, India, Japan, Pakistan, Russia, South Korea and Vietnam, are building or planning nuclear plants. But this global trend has not as yet extended to the U.S., where work on the last such facility began some 30 years ago.

If developed sensibly, nuclear power could be truly sustainable and essentially inexhaustible and

Overview: Nuclear Recycling

- To minimize global warming, humanity may need to generate much of its future energy using nuclear power technology, which itself releases essentially no carbon dioxide.

- Should many more of today's thermal (or slow-neutron) nuclear power plants be built, however, the world's reserves of low-cost uranium ore will be tapped out within several decades. In addition, large quantities of highly radioactive waste produced just in the U.S. will have to be stored for at least 10,000 years—much more than can be accommodated by the Yucca Mountain repository in Nevada. Worse, most of the energy that could be extracted from the original uranium ore would be socked away in that waste.

- The utilization of a new, much more efficient nuclear fuel cycle—one based on fast-neutron reactors and the recycling of spent fuel by pyrometallurgical processing—would allow vastly more of the energy in the earth's readily available uranium ore to be used to produce electricity. Such a cycle would greatly reduce the creation of long-lived reactor waste and could support nuclear power generation indefinitely.

could operate without contributing to climate change. In particular, a relatively new form of nuclear technology could overcome the principal drawbacks of current methods—namely, worries about reactor accidents; the potential for diversion of nuclear fuel into highly destructive weapons; the management of dangerous, long-lived radioactive waste; and the depletion of global reserves of economically available uranium. This nuclear fuel cycle would combine two innovations: pyrometallurgical processing (a high-temperature method of recycling reactor waste into fuel) and advanced fast-neutron reactors capable of burning that fuel. With this approach, the radioactivity from the generated waste could drop to safe levels in a few hundred years, thereby eliminating the need to segregate waste for tens of thousands of years.

For neutrons to cause nuclear fission efficiently, they must be traveling either very slowly or very quickly. Most existing nuclear power plants contain what are called thermal reactors, which are driven by neutrons of relatively low speed (or energy) ricocheting within their cores. Although thermal reactors generate heat and thus electricity quite efficiently, they cannot minimize the output of radioactive waste.

All reactors produce energy by splitting the nuclei of heavy-metal (high-atomic-weight) atoms, mainly uranium or elements derived from uranium. In nature, uranium occurs as a mixture of two isotopes, the easily fissionable uranium 235 (which is said to be "fissile") and the much more stable uranium 238.

The uranium fire in an atomic reactor is both ignited and sustained by neutrons. When the nucleus of a fissile atom is hit by a neutron, especially a slow-moving one, it will most likely cleave (fission), releasing substantial amounts of energy and several other neutrons. Some of these emitted neutrons then strike other nearby fissile atoms, causing them to break apart, thus propagating a nuclear chain reaction. The resulting heat is conveyed out of the reactor, where it turns water into steam that is used to run a turbine that drives an electric generator.

Uranium 238 is not fissile; it is called "fissionable" because it sometimes splits when hit by a fast neutron. It is also said to be "fertile," because when a uranium 238 atom absorbs a neutron without splitting, it transmutes into plutonium 239, which, like uranium 235, is fissile and can sustain a chain reaction. After about three years of service, when technicians typically remove used fuel from one of today's reactors because of radiation-related degradation and the depletion of the uranium 235, plutonium is contributing more than half the power the plant generates.

In a thermal reactor, the neutrons, which are born fast, are slowed (or moderated) by interactions with nearby low-atomic-weight atoms, such as the hydrogen in the water that flows through reactor cores. All but two of the 440 or so commercial nuclear reactors operating are thermal, and most of them—including the 103 U.S. power reactors—employ water both to slow neutrons and to carry fission-created heat to the associated electric generators. Most of these thermal systems are what engineers call light-water reactors.

In any nuclear power plant, heavy-metal atoms are consumed as the fuel "burns." Even though the plants begin with fuel that has had its uranium 235 content enriched, most of that easily fissioned uranium is gone after about three years. When technicians remove the depleted fuel, only about one-twentieth of the potentially fissionable atoms in it (uranium 235, plutonium and uranium 238) have been used up, so the so-called spent fuel still contains about 95 percent of its original energy. In addition, only about one-tenth of the mined uranium ore is converted into fuel in the enrichment process (during which the concentration of uranium 235 is increased considerably), so less than a hundredth of the ore's total energy content is used to generate power in today's plants.

This fact means that the used fuel from current thermal reactors still has the potential to stoke many a nuclear fire. Because the world's uranium supply is finite and the continued growth in the numbers of thermal reactors could exhaust the available low-cost uranium reserves in a few decades, it makes little sense to discard this spent fuel or the "tailings" left over from the enrichment process.

The spent fuel consists of three classes of materials. The fission products, which make up about 5 percent of the used fuel, are the true wastes—the ashes, if you will, of the fission fire. They comprise a mélange of lighter elements created when the heavy atoms split. The mix is highly radioactive for its first several years. After a decade or so, the activity is dominated by two isotopes, cesium 137 and strontium 90. Both are soluble in water, so they must be contained very securely. In around three centuries, those isotopes' radioactivity declines by a factor of 1,000, by which point they have become virtually harmless.

Uranium makes up the bulk of the spent nuclear fuel (around 94 percent); this is unfissioned uranium that has lost most of its uranium 235 and resembles natural uranium (which is just 0.71 percent fissile uranium 235). This component is only mildly radioactive and, if separated from the fission products and the rest of the material in the spent fuel, could readily be stored safely for future use in lightly protected facilities.

The balance of the material—the truly troubling part—is the transuranic component, elements heavier than uranium. This part of the fuel is mainly a blend of plutonium isotopes, with a significant presence of americium. Although the transuranic elements make up only about 1 percent of the spent fuel, they constitute the main source of today's nuclear waste problem. The half-lives (the period in which radioactivity halves) of

these atoms range up to tens of thousands of years, a feature that led U.S. government regulators to require that the planned high-level nuclear waste repository at Yucca Mountain in Nevada isolate spent fuel for over 10,000 years.

An Outdated Strategy

Early nuclear engineers expected that the plutonium in the spent fuel of thermal reactors would be removed and then used in fast-neutron reactors, called fast breeders because they were designed to produce more plutonium than they consume. Nuclear power pioneers also envisioned an energy economy that would involve open commerce in plutonium. Plutonium can be used to make bombs, however. As nuclear technology spread beyond the major superpowers, this potential application led to worries over uncontrolled proliferation of atomic weapons to other states or even to terrorist groups.

The Nuclear Non-Proliferation Treaty partially addressed that problem in 1968. States that desired the benefits of nuclear power technology could sign the treaty and promise not to acquire nuclear weapons, whereupon the weapons-holding nations agreed to assist the others with peaceful applications. Although a cadre of international inspectors has since monitored member adherence to the treaty, the effectiveness of that international agreement has been spotty because it lacks effective authority and enforcement means.

Nuclear-weapons designers require plutonium with a very high plutonium 239 isotopic content, whereas plutonium from commercial power plants usually contains substantial quantities of the other isotopes of plutonium, making it difficult to use in a bomb. Nevertheless, use of plutonium from spent fuel in weapons is not inconceivable. Hence, President Jimmy Carter banned civilian reprocessing of nuclear fuel in the U.S. in 1977. He reasoned that if plutonium were not recovered from spent fuel it could not be used to make bombs.

Carter also wanted America to set an example for the rest of the world. France, Japan, Russia and the UK have not, however, followed suit, so plutonium reprocessing for use in power plants continues in a number of nations.

An Alternative Approach

When the ban was issued, "reprocessing" was synonymous with the PUREX (for *p*lutonium *ura*nium *ex*traction) method, a technique developed to meet the need for chemically pure plutonium for atomic weapons. Advanced fast-neutron reactor technology, however, permits an alternative recycling strategy that does not involve pure plutonium at any stage. Fast reactors can thus minimize the risk that spent fuel from energy production would be used for weapons production, while providing a unique ability to squeeze the maximum energy out of nuclear fuel [see sidebar, "New Way to Reuse Nuclear Fuel," page 102]. Several such reactors have been built and used for power generation—in France, Japan, Russia, the UK and the U.S.—two of which are still operating [see "Next-Generation Nuclear Power," page 75].

Fast reactors can extract more energy from nuclear fuel than thermal reactors do because their rapidly moving (higher-energy) neutrons cause atomic fissions more efficiently than the slow thermal neutrons do. This effectiveness stems from two phenomena. At slower speeds, many more neutrons are absorbed in nonfission reactions and are lost. Second, the higher energy of a fast neutron makes it much more likely that a fertile heavy-metal atom like uranium 238 will fission when struck. Because of this fact, not only are uranium 235 and plutonium 239 likely to fission in a fast reactor, but an appreciable fraction of the heavier transuranic atoms will do so as well.

Water cannot be employed in a fast reactor to carry the heat from the core—it would slow the fast neutrons. Hence, engineers typically use a liquid metal such as sodium as a coolant and heat transporter. Liquid metal has one big advantage over water. Water-cooled systems run at very high pressure, so that a small leak can quickly develop into a large release of steam and perhaps a serious pipe break, with rapid loss of reactor coolant. Liquid-metal systems, however, operate at atmospheric pressure, so they present vastly less potential for a major release. Nevertheless, sodium catches fire if exposed to water, so it must be managed carefully. Considerable industrial experience with handling the substance has been amassed over the years, and management methods are well developed. But sodium fires have occurred, and undoubtedly there will be more. One sodium fire began in 1995 at the Monju fast reactor in Japan. It made a mess in the reactor building but never posed a threat to the integrity of the reactor, and no one was injured or irradiated. Engineers do not consider sodium's flammability to be a major problem.

Researchers at Argonne National Laboratory began developing fast-reactor technology in the 1950s. In the 1980s this research was directed toward a fast reactor (dubbed the advanced liquid-metal reactor, or ALMR), with metallic fuel cooled by a liquid metal, that was to be integrated with a high-temperature pyrometallurgical processing unit for recycling and replenishing the fuel. Nuclear engineers have also investigated several other fast-reactor concepts, some burning metallic uranium or plutonium fuels, others using oxide fuels. Coolants of liquid lead or a lead-bismuth solution have been used. Metallic fuel, as used in the ALMR, is preferable to oxide for several reasons: it has some safety advantages, it will permit faster breeding of new fuel and it can more easily be paired with pyrometallurgical recycling.

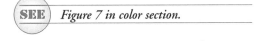 *Figure 7 in color section.*

Pyroprocessing

The pyrometallurgical process ("pyro" for short) extracts from used fuel a mix of transuranic elements instead of pure plutonium, as in the PUREX route. It is based on electroplating—using electricity to collect, on a conducting metal electrode, metal extracted as ions from a chemical bath. Its name derives from the high temperatures to which the metals must be subjected during the procedure. Two similar approaches have been developed, one in the U.S., the other in Russia. The major difference is that the Russians process ceramic (oxide) fuel, whereas the fuel in an ALMR is metallic.

In the American pyroprocess [see sidebar, "New Way to Reuse Nuclear Fuel," below], technicians dissolve spent metallic fuel in a chemical bath. Then a strong electric current selectively collects

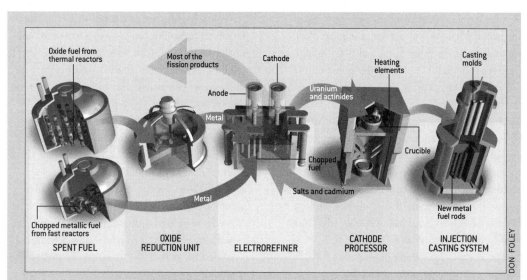

New Way to Reuse Nuclear Fuel

The key to pyrometallurgical recycling of nuclear fuel is the electrorefining procedure. This process removes the true waste, the fission products, from the uranium, the plutonium and the other actinides (heavy radioactive elements) in the spent fuel. The actinides are kept mixed with the plutonium so it cannot be used directly in weapons.

Spent fuel from today's thermal reactors (uranium and plutonium oxide) would first undergo oxide reduction to convert it to metal, whereas spent metallic uranium and plutonium fuel from fast reactors would go straight to the electrorefiner. Electrorefining resembles electroplating: spent fuel attached to an anode would be suspended in a chemical bath; then electric current would plate out uranium and other actinides on the cathode. The extracted elements would next be sent to the cathode processor to remove residual salts and cadmium from refining. Finally, the remaining uranium and actinides would be cast into fresh fuel rods, and the salts and cadmium would be recycled.

the plutonium and the other transuranic elements on an electrode, along with some of the fission products and much of the uranium. Most of the fission products and some of the uranium remain in the bath. When a full batch is amassed, operators remove the electrode. Next they scrape the accumulated materials off the electrode, melt them down, cast them into an ingot and pass the ingot to a refabrication line for conversion into fast-reactor fuel. When the bath becomes saturated with fission products, technicians clean the solvent and process the extracted fission products for permanent disposal.

Thus, unlike the current PUREX method, the pyroprocess collects virtually all the transuranic elements (including the plutonium), with considerable carryover of uranium and fission products. Only a very small portion of the transuranic component ends up in the final waste stream, which reduces the needed isolation time drastically. The combination of fission products and transuranics is unsuited for weapons or even for thermal-reactor fuel. This mixture is, however, not only tolerable but advantageous for fueling fast reactors.

Although pyrometallurgical recycling technology is not quite ready for immediate commercial use, researchers have demonstrated its basic principles. It has been successfully demonstrated on a pilot level in operating power plants, both in the U.S. and in Russia. It has not yet functioned, however, on a full production scale.

Comparing Cycles

The operating capabilities of thermal and fast reactors are similar in some ways, but in others the differences are huge [see sidebar, "Comparing Three Nuclear Fuel Cycles," page 104]. A 1,000-megawatt-electric thermal-reactor plant, for example, generates more than 100 tons of spent fuel a year. The annual waste output from a fast reactor with the same electrical capacity, in contrast, is a little more than a single ton of fission products, plus trace amounts of transuranics.

Waste management using the ALMR cycle would be greatly simplified. Because the fast-reactor waste would contain no significant quantity of long-lived transuranics, its radiation would decay to the level of the ore from which it came in several hundred years, rather than tens of thousands.

If fast reactors were used exclusively, transportation of highly radioactive materials would occur only under two circumstances—when the fission product waste was shipped to Yucca Mountain or an alternative site for disposal and when start-up fuel was shipped to a new reactor. Commerce in plutonium would be effectively eliminated.

Some people are advocating that the U.S. embark on an extensive program of PUREX processing of reactor fuel, making mixed oxides of uranium and plutonium for cycling back into thermal reactors. Although the mixed oxide (MOX) method is currently being used for spoiling excess weapons plutonium so that it cannot be employed in bombs—a good idea—we think that it would be a mistake to deploy the much larger PUREX infrastructure that would be required to process civilian fuel. The resource gains would be modest, whereas the long-term waste problem would remain, and the entire effort would delay for only a short time the need for efficient fast reactors.

The fast-reactor system with pyroprocessing is remarkably versatile. It could be a net consumer or net producer of plutonium, or it could be run in a break-even mode. Operated as a net producer, the system could provide start-up materials for other fast-reactor power plants. As a net consumer, it could use up excess plutonium and weapons materials. If a break-even mode were chosen, the only additional fuel a nuclear plant would need would be a periodic infusion of depleted uranium (uranium from which most of the fissile uranium 235

Comparing Three Nuclear Fuel Cycles

*Three major approaches to burning nuclear fuel and handling its wastes can be employed;
some of their features are noted below.*

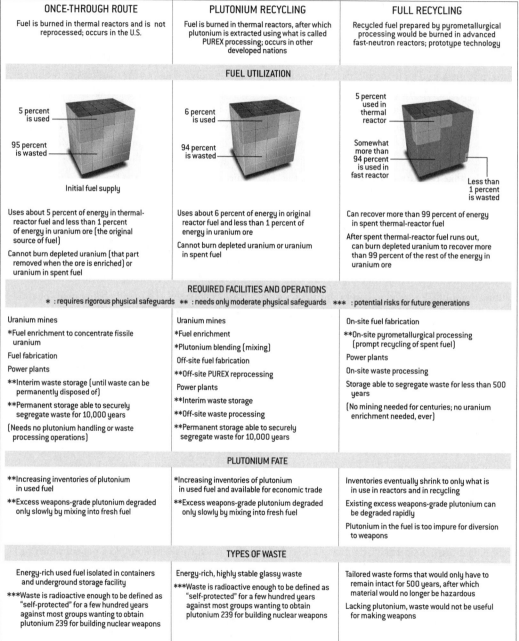

ONCE-THROUGH ROUTE	PLUTONIUM RECYCLING	FULL RECYCLING
Fuel is burned in thermal reactors and is not reprocessed; occurs in the U.S.	Fuel is burned in thermal reactors, after which plutonium is extracted using what is called PUREX processing; occurs in other developed nations	Recycled fuel prepared by pyrometallurgical processing would be burned in advanced fast-neutron reactors; prototype technology

FUEL UTILIZATION

5 percent is used 95 percent is wasted *Initial fuel supply*	6 percent is used 94 percent is wasted	5 percent used in thermal reactor Somewhat more than 94 percent is used in fast reactor Less than 1 percent is wasted
Uses about 5 percent of energy in thermal-reactor fuel and less than 1 percent of energy in uranium ore (the original source of fuel) Cannot burn depleted uranium (that part removed when the ore is enriched) or uranium in spent fuel	Uses about 6 percent of energy in original reactor fuel and less than 1 percent of energy in uranium ore Cannot burn depleted uranium or uranium in spent fuel	Can recover more than 99 percent of energy in spent thermal-reactor fuel After spent thermal-reactor fuel runs out, can burn depleted uranium to recover more than 99 percent of the rest of the energy in uranium ore

REQUIRED FACILITIES AND OPERATIONS

* : requires rigorous physical safeguards ** : needs only moderate physical safeguards *** : potential risks for future generations

Uranium mines *Fuel enrichment to concentrate fissile uranium Fuel fabrication Power plants **Interim waste storage (until waste can be permanently disposed of) **Permanent storage able to securely segregate waste for 10,000 years (Needs no plutonium handling or waste processing operations)	Uranium mines *Fuel enrichment *Plutonium blending (mixing) Off-site fuel fabrication **Off-site PUREX reprocessing Power plants **Interim waste storage **Off-site waste processing **Permanent storage able to securely segregate waste for 10,000 years	On-site fuel fabrication **On-site pyrometallurgical processing (prompt recycling of spent fuel) Power plants On-site waste processing Storage able to segregate waste for less than 500 years (No mining needed for centuries; no uranium enrichment needed, ever)

PLUTONIUM FATE

**Increasing inventories of plutonium in used fuel **Excess weapons-grade plutonium degraded only slowly by mixing into fresh fuel	*Increasing inventories of plutonium in used fuel and available for economic trade **Excess weapons-grade plutonium degraded only slowly by mixing into fresh fuel	Inventories eventually shrink to only what is in use in reactors and in recycling Existing excess weapons-grade plutonium can be degraded rapidly Plutonium in the fuel is too impure for diversion to weapons

TYPES OF WASTE

Energy-rich used fuel isolated in containers and underground storage facility ***Waste is radioactive enough to be defined as "self-protected" for a few hundred years against most groups wanting to obtain plutonium 239 for building nuclear weapons	Energy-rich, highly stable glassy waste ***Waste is radioactive enough to be defined as "self-protected" for a few hundred years against most groups wanting to obtain plutonium 239 for building nuclear weapons	Tailored waste forms that would only have to remain intact for 500 years, after which material would no longer be hazardous Lacking plutonium, waste would not be useful for making weapons

DON FOLEY

has been removed) to replace the heavy-metal atoms that have undergone fission.

Business studies have indicated that this technology could be economically competitive with existing nuclear power technologies. Certainly, pyrometallurgical recycling will be dramatically less expensive than PUREX reprocessing, but in truth, the economic viability of the system cannot be known until it is demonstrated.

The overall economics of any energy source depend not only on direct costs but also on what economists call "externalities," the hard-to-quantify costs of outside effects resulting from using the technology. When we burn coal or oil to make electricity, for example, our society accepts the detrimental health effects and the environmental costs they entail. Thus, external costs in effect subsidize fossil fuel power generation, either directly or via indirect effects on the society as a whole. Even though they are difficult to reckon, economic comparisons that do not take externalities into account are unrealistic and misleading.

Coupling Reactor Types

If advanced fast reactors come into use, they will at first burn spent thermal-reactor fuel that has been recycled using pyroprocessing. That waste, which is now "temporarily" stored on site, would be transported to plants that could process it into three output streams. The first, highly radioactive stream would contain most of the fission products, along with unavoidable traces of transuranic elements. It would be transformed into a physically stable form—perhaps a glasslike substance—and then shipped to Yucca Mountain or some other permanent disposal site.

The second stream would capture virtually all the transuranics, together with some uranium and fission products. It would be converted to a metallic fast-reactor fuel and then transferred to ALMR-type reactors.

The third stream, amounting to about 92 percent of the spent thermal-reactor fuel, would contain the bulk of the uranium, now in a depleted state. It could be stashed away for future use as fast-reactor fuel.

Such a scenario cannot be realized overnight, of course. If we were to begin today, the first of the fast reactors might come online in about 15 years. Notably, that schedule is reasonably compatible with the planned timetable for shipment of spent thermal-reactor fuel to Yucca Mountain. It could instead be sent for recycling into fast-reactor fuel.

As today's thermal reactors reach the end of their lifetimes, they could be replaced by fast reactors. Should that occur, there would be no need to mine any more uranium ore for centuries and no further requirement, ever, for uranium enrichment. For the very long term, recycling the fuel of fast reactors would be so efficient that currently available uranium supplies could last indefinitely.

Both India and China have recently announced that they plan to extend their energy resources by deploying fast reactors. We understand that their first fast reactors will use oxide or carbide fuel rather than metal—a less than optimum path, chosen presumably because the PUREX reprocessing technology is mature, whereas pyroprocessing has not yet been commercially demonstrated.

It is not too soon for the U.S. to complete the basic development of the fast-reactor/pyroprocessing system for metallic fuel. For the foreseeable future, the hard truth is this: only nuclear power can satisfy humanity's long-term energy needs while preserving the environment. For large-scale, sustainable nuclear energy production to continue, the supply of nuclear fuel must last a long time. That means that the nuclear power cycle must have the characteristics of the ALMR and pyroprocessing. The time seems right to take this new course toward sensible energy development.

—DECEMBER 2005

The Nuclear Option

A threefold expansion of nuclear power could contribute significantly to staving off climate change by avoiding one billion to two billion tons of carbon emissions annually.

JOHN M. DEUTCH AND ERNEST J. MONIZ

Nuclear power supplies a sixth of the world's electricity. Along with hydropower (which supplies slightly more than a sixth), it is the major source of "carbon-free" energy today. The technology suffered growing pains, seared into the public's mind by the Chernobyl and Three Mile Island accidents, but plants have demonstrated remarkable reliability and efficiency recently. The world's ample supply of uranium could fuel a much larger fleet of reactors than exists today throughout their 40- to 50-year life span.

With growing worries about global warming and the associated likelihood that greenhouse gas emissions will be regulated in some fashion, it is not surprising that governments and power

Overview

- Global electricity consumption is projected to increase 160 percent by 2050.
- Building hundreds of nuclear power plants will help meet that need without large new emissions of carbon dioxide.
- This scenario requires economical new plants, a plan for waste storage and prevention of nuclear weapons proliferation.

providers in the U.S. and elsewhere are increasingly considering building a substantial number of additional nuclear power plants. The fossil fuel alternatives have their drawbacks. Natural gas is attractive in a carbon-constrained world because it has lower carbon content relative to other fossil fuels and because advanced power plants have low capital costs. But the cost of the electricity produced is very sensitive to natural gas prices, which have become much higher and more volatile in recent years. In contrast, coal prices are relatively low and stable, but coal is the most carbon-intensive source of electricity. The capture and sequestration of carbon dioxide, which will add significantly to the cost, must be demonstrated and introduced on a large scale if coal-powered electricity is to expand significantly without emitting unacceptable quantities of carbon into the atmosphere. These concerns raise doubts about new investments in gas- or coal-powered plants.

All of which points to a possible nuclear revival. And indeed, more than 20,000 megawatts of nuclear capacity have come online globally since 2000, mostly in the Far East. Yet despite the evident interest among major nuclear operators, no firm orders have been placed in the U.S. Key impediments to new nuclear construction are high capital costs and the uncertainty surrounding nuclear waste management. In addition, global expansion of nuclear power has raised concerns that nuclear weapons ambitions in certain countries may inadvertently be advanced.

In 2003 we cochaired a major Massachusetts Institute of Technology study, *The Future of Nuclear Power*, that analyzed what would be required to retain the nuclear option. That study described a scenario whereby worldwide nuclear power generation could triple to one million megawatts by the year 2050, saving the globe from emissions of between 0.8 billion and 1.8 billion tons of carbon a year, depending on whether gas- or coal-powered plants were displaced. At this scale, nuclear power would significantly contribute to the stabilization of greenhouse gas emissions, which requires about seven billion tons of carbon to be averted annually by 2050 [see "A Plan to Keep Carbon in Check," page 52].

The Fuel Cycle

If nuclear power is to expand by such an extent, what kind of nuclear plants should be built? A chief consideration is the fuel cycle, which can be either open or closed. In an open fuel cycle, also known as a once-through cycle, the uranium is "burned" once in a reactor, and spent fuel is stored in geologic repositories. The spent fuel includes plutonium that could be chemically extracted and turned into fuel for use in another nuclear plant. Doing that results in a closed fuel cycle, which some people advocate [see "Smarter Use of Nuclear Waste," page 98].

Some countries, most notably France, currently use a closed fuel cycle in which plutonium is separated from the spent fuel and a mixture of plutonium and uranium oxides is subsequently burned again. A longer-term option could involve recycling all the transuranics (plutonium is one example of a transuranic element), perhaps in a so-called fast reactor. In this approach, nearly all the very long-lived components of the waste are eliminated, thereby transforming the nuclear waste debate. Substantial research and development is needed, however, to work through daunting technical and economic challenges to making this scheme work.

Recycling waste for reuse in a closed cycle might seem like a no-brainer: less raw material is used for the same total power output, and the problem of long-term storage of waste is alleviated because a smaller amount of radioactive material must be stored for many thousands of years. Nevertheless, we believe that an open cycle is to be preferred over the next several decades. First, the recycled fuel is more expensive than the original uranium. Second, there appears to be ample uranium at reasonable cost to sustain the tripling in global nuclear power generation that we envisage with a once-through fuel cycle for the entire lifetime of the nuclear fleet (about 40 to 50 years for each plant). Third, the environmental benefit for long-term waste storage is offset by near-term risks to the environment from the complex and highly dangerous reprocessing and fuel fabrication operations. Finally, the reprocessing that occurs in a closed fuel cycle produces plutonium that can be diverted for use in nuclear weapons.

The type of reactor that will continue to dominate for at least two decades, probably longer, is the light-water reactor, which uses ordinary water (as opposed to heavy water, containing deuterium) as the coolant and moderator. The vast majority of plants in operation in the world today are of this type, making it a mature, well-understood technology.

Reactor designs are divided into generations. The earliest prototype reactors, built in the 1950s and early 1960s, were often one of a kind. Generation II reactors, in contrast, were commercial designs built in large numbers from the late 1960s to the early 1990s. Generation III reactors incorporate design improvements such as better fuel technology and passive safety, meaning that in

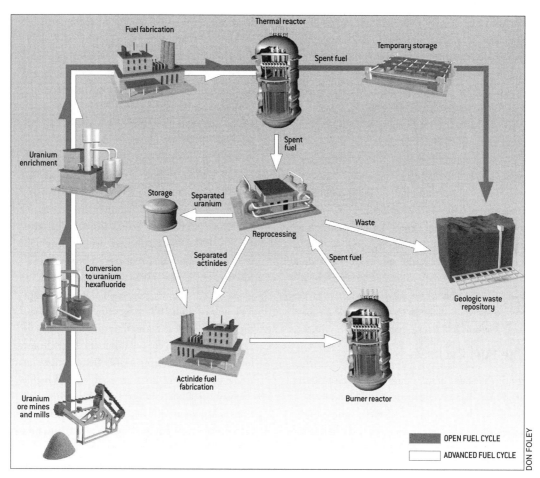

Fuel fabrication

Thermal reactor

Temporary storage

Spent fuel

Uranium enrichment

Storage

Separated uranium

Spent fuel

Reprocessing

Waste

Conversion to uranium hexafluoride

Separated actinides

Spent fuel

Geologic waste repository

Uranium ore mines and mills

Actinide fuel fabrication

Burner reactor

OPEN FUEL CYCLE

ADVANCED FUEL CYCLE

DON FOLEY

PREFERRED FUEL CYCLES

The authors prefer an open fuel cycle for the next several decades: the uranium is burned once in a thermal reactor, and the spent fuel is stored in a waste repository (dark path). Some countries currently use a closed cycle in which plutonium is extracted from spent fuel and mixed with uranium for reuse in a thermal reactor (not shown). An advanced closed cycle (white path) might become feasible and preferred in the distant future: plutonium and other elements (actinides) and perhaps the uranium in spent fuel would be reprocessed and used in special burner reactors, dramatically reducing the quantity of waste requiring long-term storage.

the case of an accident the reactor shuts itself down without requiring the operators to intervene. The first Generation III reactor was built in Japan in 1996. Generation IV reactors are new designs that are currently being researched, such as pebble-bed reactors and lead-cooled fast reactors [see

"Next-Generation Nuclear Power," page 75]. In addition, Generation III+ reactors are designs similar to Generation III but with the advanced features further evolved. With the possible exception of high-temperature gas reactors (the pebble bed is one example), Generation IV reactors are several decades away from being candidates for significant commercial deployment. To evaluate our scenario through to 2050, we envisaged the building of Generation III+ light-water reactors.

The pebble-bed modular reactor introduces the interesting prospect of modular nuclear plants. Instead of building a massive 1,000-megawatt plant, modules each producing around 100 megawatts can be built. This approach may be particularly attractive, both in developing countries and in deregulated industrial countries, because of the much lower capital costs involved. The traditional large plants do have the advantage of economy of scale, most likely resulting in lower cost per kilowatt of capacity, but this edge could be challenged if efficient factory-style production of large numbers of modules could be implemented. South Africa is scheduled to begin construction of a 110-megawatt demonstration pebble-bed plant in 2007, to be completed by 2011, with commercial modules of about 165 megawatts planned for 2013. The hope is to sell modules internationally, in particular throughout Africa.

Reducing Costs

Based on previous experience, electricity from new nuclear power plants is currently more expensive than that from new coal- or gas-powered plants. The 2003 MIT study estimated that new light-water reactors would produce electricity at a cost of 6.7 cents per kilowatt-hour. That figure includes all the costs of a plant, spread over its life span, and includes items such as an acceptable return to investors. In comparison, under equivalent assumptions we estimated that a new coal plant would produce electricity at a cost of 4.2 cents per kilowatt-hour. For a new gas-powered plant, the cost is very sensitive to the price of natural gas and would be about 5.8 cents per kilowatt-hour for today's high gas prices (about $7 per million Btu).

Some people will be skeptical about how well the cost of nuclear power can be estimated, given past overoptimism, going back to claims in the early days that nuclear power would be "too cheap to meter." But the MIT analysis is grounded in past experience and actual performance of existing plants, not in promises from the nuclear industry. Some might also question the uncertainties inherent in such cost projections. The important point is that the estimates place the three alternatives—nuclear, coal and gas—on a level playing field, and there is no reason to expect unanticipated contingencies to favor one over the other. Furthermore, when utilities are deciding what kind of power plant to build, they will base their decisions on such estimates.

Several steps could reduce the cost of the nuclear option below our baseline figure of 6.7 cents per kilowatt-hour. A 25 percent reduction in construction expenses would bring the cost of electricity down to 5.5 cents per kilowatt-hour. Reducing the construction time of a plant from five to four years and improvements in operation and maintenance can shave off a further 0.4 cent per kilowatt-hour. How any plant is financed can depend dramatically on what regulations govern the plant site. Reducing the cost of capital for a nuclear plant to be the same as for a gas or coal plant would close the gap with coal (4.2 cents per kilowatt-hour). All these reductions in the cost of nuclear power are plausible—particularly if the industry builds a large number of just a few standardized designs—but not yet proved.

Nuclear power becomes distinctly favored economically if carbon emissions are priced [see illustrations on page 110]. We will refer to this as a carbon tax, but the pricing mechanism need not be in the form of a tax. Europe has a system in which permits to emit carbon are traded on an open market. In early 2006 permits were selling for more than $100 per tonne of carbon emitted (or $27 per tonne of carbon dioxide), although

Into the Future

The global demand for electricity will increase substantially in the coming decades. To meet that demand, thousands of new power plants must be built. One of the most significant factors in determining what kind of facilities are built will be the estimated cost of the electricity produced (right). Nuclear plants will not be built in large numbers if they are not economically competitive with coal- and gas-powered plants. If nuclear plants can be made competitive, global nuclear power production might triple from 2000 to 2050, a scenario evaluated by an MIT study (below).

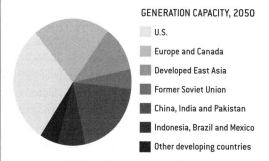

GENERATION CAPACITY, 2050

- U.S.
- Europe and Canada
- Developed East Asia
- Former Soviet Union
- China, India and Pakistan
- Indonesia, Brazil and Mexico
- Other developing countries

Who will have the power?

The MIT scenario projects that the U.S. will produce about a third of the one million megawatts of electricity that will be generated by nuclear power in 2050 and that the rest of the developed world will provide another third.

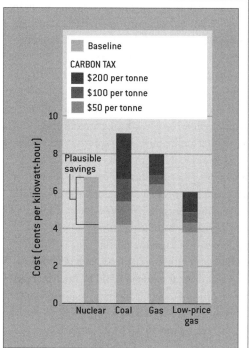

Paying the Piper

The cost of electricity projected for newly built power plants depends on many factors. Taxes on carbon emissions could raise costs for coal and gas. Nuclear may be reduced by plausible but unproved cost-cutting steps.

JEN CHRISTIANSEN, SOURCES: ENERGY INFORMATION ADMINISTRATION AND ERNEST J. MONIZ (ELECTRICITY CONSUMPTION); "THE FUTURE OF NUCLEAR POWER," BY STEPHEN ANSOLABEHERE ET AL., MASSACHUSETTS INSTITUTE OF TECHNOLOGY, 2003 (COST OF ELECTRICITY AND GENERATION CAPACITY)

recently their price has fallen to about half that. (A metric unit, one tonne is equal to 1.1 U.S. tons.) A tax of only $50 per tonne of carbon raises coal-powered electricity to 5.4 cents per kilowatt-hour. At $200 per tonne of carbon, coal reaches a whopping 9.0 cents per kilowatt-hour. Gas fares much better than coal, increasing to 7.9 cents per kilowatt-hour under a $200 tax. Fossil fuel plants could avoid the putative carbon tax by capturing and sequestering the carbon, but the cost of doing that contributes in the same way that a tax would [see "Can We Bury Global Warming?" page 44].

Because it is many years since construction of a nuclear plant was embarked on in the U.S., the companies that build the first few new plants will face extra expenses that subsequent operators will not have to bear, along with additional risk in working through a new licensing process. To help overcome that hurdle, the Energy Policy Act of 2005 included a number of important provisions, such as a tax credit of 1.8 cents per kilowatt-hour to new nuclear plants for their first eight years of operation. The credit, sometimes called a first-mover incentive, applies to the first 6,000 megawatts of new plants to come online. Several consortiums have formed to take advantage of the new incentives.

Waste Management

The second big obstacle that a nuclear renaissance faces is the problem of waste management. No country in the world has yet implemented a system for permanently disposing of the spent fuel and other radioactive waste produced by nuclear power plants. The most widely favored approach is geologic disposal, in which waste is stored in chambers hundreds of meters underground. The goal is to prevent leakage of the waste for many millennia through a combination of engineered barriers (for example, the waste containers) and geologic ones (the natural rock structure where the chamber has been excavated and the favorable characteristics of the hydrogeologic basin). Decades of studies support the geologic disposal option. Scientists have a good understanding of the processes and events that could transport radionuclides from the repository to the biosphere. Despite this scientific confidence, the process of approving a geologic site remains fraught with difficulties.

A prime case in point is the proposed facility at Yucca Mountain in Nevada, which has been under consideration for two decades. Recently the site was found to have considerably more water than anticipated. It remains uncertain whether the Nuclear Regulatory Commission (NRC) will license the site.

Delays in resolving waste management (even if it is approved, it is unlikely that Yucca Mountain will be accepting waste before 2015) may complicate efforts to construct new power plants. By law, the government was to begin moving spent fuel from reactor sites to a repository by 1998. Failure to do so has led to a need for increased local storage at many sites and associated unhappiness among neighbors, towns and states.

Perhaps the first country to build a permanent storage site for its high-level nuclear waste will be Finland. At Olkiluoto, the location of two nuclear reactors, excavation has begun on an underground research facility called Onkalo. Extending about half a kilometer underground, the Onkalo project will involve study of the rock structure and groundwater flows and will test the disposal technology in actual deep underground conditions. If all goes according to plan and the necessary government licenses are obtained, the first canisters of waste could be emplaced in 2020. By 2130 the repository would be complete, and the access routes would be filled and sealed. The money to pay for the facility has been levied on the price of Finnish nuclear power since the late 1970s.

To address the waste management problem in the U.S., the government should take title to the spent fuel stored at commercial reactor sites across the country and consolidate it at one or more federal interim storage sites until a permanent disposal facility is built. The waste can be temporarily stored safely and securely for an extended period. Such extended temporary storage, perhaps even for

Nuclear Waste Disposal

Finland is moving ahead with a project to investigate underground disposal of nuclear waste at Olkiluoto. Under the plan, spent fuel rods will be encapsulated in large canisters made of an inner shell of iron for mechanical strength and a thick outer shell of copper to resist corrosion. The canisters will be placed in holes bored into the tunnel floors and surrounded by clay to prevent direct water flow to the canisters. The facility is to begin accepting waste from Finland's four nuclear reactors in 2020.

as long as 100 years, should be an integral part of the disposal strategy. Among other benefits, it would take the pressure off government and industry to come up with a hasty disposal solution.

Meanwhile the Department of Energy should not abandon Yucca Mountain. Instead it should reassess the suitability of the site under various conditions and modify the project's schedule as needed. If nuclear power expanded globally to one million megawatts, enough high-level waste and spent fuel would be generated in the open fuel cycle to fill a Yucca Mountain–size facility every three and a half years. In the court of public opinion, that fact is a significant disincentive to the expansion of nuclear power, yet it is a problem that can and must be solved.

The Threat of Proliferation

In conjunction with the domestic program of waste management just outlined, the president should continue the diplomatic effort to create an international system of fuel supplier countries and user countries. Supplier countries such as the U.S., Russia, France and the UK would sell fresh fuel to user countries with smaller nuclear programs and commit to removing the spent fuel from them. In return, the user countries would forgo the construction of fuel-producing facilities. This arrangement would greatly alleviate the danger of nuclear weapons proliferation because the chief risks for proliferation involve not the nuclear power plants themselves but the fuel enrichment and reprocessing plants. The current situation with Iran's uranium enrichment program is a prime example. A scheme in which fuel is leased to users is a necessity in a world where nuclear power is to expand threefold, because such an expansion will inevitably involve the spread of nuclear power plants to some countries of proliferation concern.

A key to making the approach work is that producing fuel does not make economic sense for small nuclear power programs. This fact underlies the marketplace reality that the world is already divided into supplier and user countries. Instituting the supplier/user model is largely a matter, albeit not a simple one, of formalizing the current situation more permanently through new agreements that reinforce commercial realities.

Although the proposed regime is inherently attractive to user nations—they get an assured supply of cheap fuel and are relieved of the problem of dealing with waste materials—other incentives should also be put in place because the user states would be agreeing to go beyond the requirements of the treaty on the nonproliferation of nuclear weapons. For example, if a global system of tradable carbon credits were instituted, user nations adhering to the fuel-leasing rules could be granted credits for their new nuclear power plants.

Iran is the most obvious example today of a nation that the global community would rather see as a "user state" than as a producer of enriched uranium. But it is not the only difficult case. Another nation whose program must be addressed promptly is Brazil, where an enrichment facility is under construction supposedly to provide fuel for the country's two nuclear reactors. A consistent approach to countries such as Iran and Brazil will be needed if nuclear power is to be expanded globally without exacerbating proliferation concerns.

The Terawatt Future

A terawatt—one million megawatts—of "carbon-free" power is the scale needed to make a significant dent in projected carbon dioxide emissions at midcentury. In the terms used by Socolow and Pacala [see "A Plan to Keep Carbon in Check," page 52], that contribution would correspond to one to two of the seven required "stabilization wedges." Reaching a terawatt of nuclear power by 2050 is certainly challenging, requiring deployment of about 2,000 megawatts a month. A capital investment of $2 trillion over several decades is called for, and power plant cost reduction, nuclear waste management and a proliferation-resistant international fuel cycle regime must all be addressed aggressively over the next decade or so. A critical determinant will be the degree to which carbon dioxide emissions from fossil fuel use are priced, both in the industrial world and in the large emerging economies such as China, India and Brazil.

The economics of nuclear power are not the only factor governing its future use. Public acceptance also turns on issues of safety and nuclear waste, and the future of nuclear power in the U.S. and much of Europe remains in question. Regarding safety, it is essential that NRC regulations be enforced diligently, which has not always been the case.

In the scenario developed as part of the MIT study, it emerged that the U.S. would approximately triple its nuclear deployment—to about 300,000 megawatts—if a terawatt were to be realized globally. The credibility of such a scenario will be largely determined in the forthcoming decade by the degree to which the first-mover incentives in the 2005 Energy Policy Act are exercised, by the capability of the government to start moving spent fuel from reactor sites and by whether the American political process results in a climate change policy that will significantly limit carbon dioxide emissions.

—SEPTEMBER 2006

FUEL CELLS AND
A HYDROGEN ECONOMY

The Power Plant in Your Basement

In the past, stationary fuel cells were megawatt behemoths, designed for the electric utilities. Now they are being shrunk for homes and other modest applications.

ALAN C. LLOYD

As deregulation of the electric utility industry dissolves the monopoly once held by most power generators, one repercussion has been increasingly long distances between some buyers and sellers of electricity. Nevertheless, within a decade or two, some customers may find themselves living in a home whose electricity comes not from a generating plant tens, hundreds or even thousands of kilometers away but rather from a refrigerator-size power station right in their own basements or backyards. Moreover, not just homes but shops, small businesses, hotels, apartment buildings and possibly factories may all be powered in the same

way: by fuel cells in the range of five to 500 kilowatts.

Companies and industrial research laboratories in Belgium, Canada, Denmark, Germany, Italy, Japan, Korea and the U.S. have aggressive fuel-cell development efforts under way, and at least a few are already selling the units. In fact, a subsidiary of United Technologies has been offering fuel cells of up to 200 kilowatts for almost a decade. They have sold about 170 units, many of which are used for generation of both heat and power at industrial facilities or for backup power. They are also increasingly being used at wastewater treatment plants and in "green" facilities, which showcase environmentally sensitive technologies and design.

At present, the high cost of fuel cells has limited their use to these and very few other special-ized applications, made feasible for the most part by generous government subsidies. Electricity from fuel cells now costs $3,000 to $4,000 per kilowatt, as opposed to $500 to $1,000 per kilowatt for the

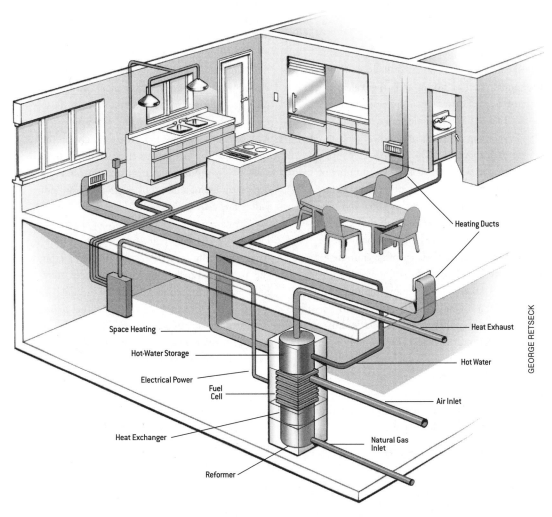

Solid-Oxide Fuel Cell

Solid-oxide fuel cell could provide electricity, heat and hot water to a home. The device operates at 800 degrees Celsius (1,500 degrees Fahrenheit), and some of the heat necessary to sustain such a temperature could be captured and directed into the home's heating ducts and into the hot-water tank. This use of heat that would otherwise be wasted enables the system to put as much as 90 percent of the fuel's chemical energy to productive use. Such a unit, which would produce up to 10 kilowatts of electricity, is being designed by Hydrogen Burner Technology in Long Beach, Calif.

ordinary gas-fired combustion turbine commonly used by utilities. Another drawback is limited lifetimes; so far no commercial fuel cell has been in operation for more than 10 years, and utilities expect to get at least 20 years of useful service life from their generating equipment.

On the other hand, fuel cells have several very desirable features: they operate relatively cleanly and silently, can use a variety of fuels, and are generally unaffected by storms and other calamities. Because of these advantages, some observers believe fuel cells can become viable for a reasonably large group of applications when per-kilowatt prices reach about $1,500.

Developers will have to achieve a number of design and manufacturing improvements before fuel cells attain even that level of price and performance. The incentives for them to do so, however, are great. As concern mounts about the harmful environmental effects of greenhouse gases from conventional power plants, increasing use of fuel cells is expected to help move industrial societies toward a "hydrogen economy." Electricity will come mainly from fuel cells and other hydrogen-based devices and from solar cells, windmills and other renewable sources—which will also electrolyze water to contribute hydrogen for the fuel cells. The shift to this hydrogen-based energy infrastructure will accelerate in coming decades, particularly as oil supplies begin dwindling.

Have Fuel, Will Energize

Fuel cells are not new; in fact, the basic concept is well over a century old. Like batteries, fuel cells come in a variety of different types. Also like batteries, they produce an electric current by intercepting the electrons that flow from one reactant to the other in an electrochemical reaction. A fuel cell consists of a positive and a negative electrode separated by an electrolyte, a material that allows the passage of charged atoms, called ions.

In operation, hydrogen is passed over the negative electrode, while oxygen is passed over the positive electrode. At the negative electrode, a highly conductive catalyst, such as platinum, strips an electron from each hydrogen atom, ionizing it. The hydrogen ion and the electron then take separate paths to the positive electrode: the hydrogen ion migrates through the electrolyte, while the electron travels on an external circuit. Along the way, these electrons can be used to power an electrical device, such as a lighting fixture or a motor. At the positive electrode, the hydrogen ions and electrons combine with oxygen to form water. (Interestingly, on the space shuttle, which gets its electricity from fuel cells, the by-product water is used for drinking.) To generate a useful amount of electric current, individual fuel cells are "stacked," like a club sandwich.

The device provides direct-current electricity as long as it is fed with hydrogen and oxygen. The oxygen typically comes from the ambient air, but the hydrogen usually comes from a system called a reformer, which produces the gas by breaking down a fossil fuel. One of the advantages of fuel cells is the great diversity of sources of suitable fuel: any hydrogen-rich material is a possible source of hydrogen. Candidates include ammonia, fossil fuels—natural gas, petroleum distillates, liquid propane and gasified coal—and renewable fuels, such as ethanol, methanol and biomass (essentially any kind of plant matter). Hydrogen can also be produced by solar, wind or geothermal plants. Even waste gases from landfills and water treatment plants will do. Reformers do release pollutants as they break down the fuel to make hydrogen. In comparison with a conventional gas-fired combustion turbine, however, the emissions are considerably less—typically a tenth to a thousandth, depending on the specific pollutant and how the emissions are controlled on the turbine.

Because a fuel cell's output is direct current, a device called an inverter is necessary to convert the

DC into alternating current before the electricity can be of any practical residential or commercial use. In both the inverter and the reformer, power is lost, mostly as heat. Thus, although fuel cells themselves can have fuel-to-electricity efficiencies in excess of 45 percent, the energy losses in the reformer and inverter can bring the overall efficiency down to approximately 40 percent—about the same as a state-of-the-art gas-fired combustion turbine. As with the combustion turbine, however, recovering the waste heat—for example, to heat water or air— boosts the efficiency of the device significantly.

A popular misconception about fuel cells intended for stationary uses is that all of them are much more powerful than those being developed to propel automobiles. In fact, a fuel cell of just 40 to 50 kilowatts can easily meet the electrical needs of a large, four- or five-bedroom house or a small commercial establishment, such as a laundry. In comparison, because of the high level of power needed to accelerate a full-size car with four passengers, a fuel cell for a vehicle generally needs to put out at least 50 kilowatts. The more demanding requirements for automotive cells have prompted some observers to speculate that in the future some rural dwellers may even get power by simply plugging their homes into their cars.

The misconception about stationary fuel cells being uniformly larger than ones for automobiles probably stems from some very large experimental units tested by electric utilities over the past 20 years. The most notable of these were a 4.5-megawatt fuel cell installed by Consolidated Edison in New York City in 1982, a 4.5-megawatt unit operated by Tokyo Electric Power in 1984, an 11-megawatt unit operated by the same company from 1991 to 1997 and a two-megawatt plant tested by Pacific Gas and Electric in Santa Clara, California, in 1995. The U.S. demonstrations were rather problematic; the northern California cell, for example, rarely generated more than one megawatt—only

half the capacity it was designed for—and the New York City cell never operated at all. The Japanese experiences, however, were much more favorable; the 11-megawatt unit, for example, ran for approximately 23,000 hours.

New Paradigm

Partly because of those difficulties, developers and proponents of stationary fuel cells have shifted to a paradigm based on a more decentralized approach. Smaller units, of less than 50 kilowatts, will supply power to individual homes, and larger systems of up to several hundred kilowatts will power commercial buildings and other enterprises. Industry sources estimate that sales of the smaller fuel cells for residences and small businesses could reach $50 billion a year in the U.S. by 2030.

Such a figure may represent a certain amount of wishful thinking. There are no single-family residences at present that receive their power from onsite fuel cells; however, three companies—Plug Power in Latham, New York; Avista Laboratories in Spokane, Washington; and Northwest Power Systems in Bend, Oregon—have units that provide electricity to demonstration homes. The first fuel cell installed permanently at a home in the U.S., a brick, ranch-style house near Albany, New York, went into operation a little over a year ago, in June 1998.

Larger systems aimed at industrial or commercial uses are also in the works. At least one company hopes to introduce a 500-kilowatt fuel cell in the next few years for stationary applications, and several others are developing or selling fuel cells rated at 200 to 250 kilowatts. A 250-kilowatt cell could supply, for example, several stores in a strip mall or a small medical or corporate center.

Where larger loads are expected, multiple units can be linked. The developers of a building recently dedicated at 4 Times Square in New York City have installed two 200-kilowatt fuel cells to provide hot water for the building, light its facade and supply backup power. The edifice is known as the Green Building because its developer, the Durst Organization, designed it partly to highlight technologies that are considered ecologically sound.

In a number of recent applications, fuel cells were chosen because their unusual features outweighed their high cost. For instance, a 200-kilowatt unit was installed at the police substation in New York City's Central Park. Use of a fuel cell obviated the expensive need to dig up the park to install underground power lines. In Nebraska the First National Bank of Omaha disclosed this past February that it would install four 200-kilowatt fuel cells at its Technology Center, where it processes credit-card transactions. The company chose fuel cells, backed up by auxiliary generators

and by conventional electrical service from the local grid, because it needed extraordinarily high reliability for this application, in which even brief interruptions are quite costly.

Five Types of Cell

Of course, stationary fuel cells can be much larger and heavier than their mobile counterparts. So this market, though tiny today, has an unusual diversity of technologies being developed or sold. There are five main types of cell, each named after the electrolyte used in the system: phosphoric acid, molten carbonate, solid oxide, proton-exchange membrane and alkaline.

The phosphoric-acid fuel cell (PAFC) is the most mature technology of the five and the only

Fuel Cells Compared

ELECTROLYTE	Proton-exchange membrane	Phosphoric acid	Molten carbonate	Solid-oxide ceramic
OPERATING TEMPERATURE	80° Celsius	Around 200° C	650° C	800–1,000° C
CHARGE CARRIER	Hydrogen ion	Hydrogen ion	Carbonate ion	Oxygen ion
REFORMER	External	External	Internal or external	Internal or external
PRIME CELL COMPONENTS	Carbon based	Graphite based	Stainless steel	Ceramic
CATALYST	Platinum	Platinum	Nickel	Perovskites (titanate of calcium)
EFFICIENCY (percent)	40 to 50	40 to 50	Greater than 60	Greater than 60
STATUS OF DEVELOPMENT	Demonstration systems up to 50 kilowatts; 250-kilowatt units expected in next few years	Commercial systems operating, most of them 200-kilowatt; an 11-megawatt model has been tested	Demonstration systems up to 2 megawatts	Units up to 100 kilowatts have been demonstrated

RICHARD HUNT

one being offered commercially as of this writing in capacities above 100 kilowatts (all the fuel cells sold so far for commercial uses are PAFCs). Around the world, 12 organizations (seven in the U.S.) are marketing PAFCs or developing them. One of the largest is ONSI, a subsidiary of United Technologies, which has been deploying the units since the late 1980s. To date, the company has installed about 170 units, almost all of which are operating on natural gas. Some of these units have operated for tens of thousands of hours.

In the U.S., many of the purchases of ONSI fuel cells were subsidized under a program run by the Department of Defense and the Department of Energy since 1996. Buyers receive $1,000 per kilowatt or a third of the total project cost, whichever is lower. Already the program has distributed more than $18 million to the buyers and installers of more than 90 fuel-cell power plants.

A few large Japanese companies have sold approximately 120 PAFCs, with capacities ranging from 50 to 500 kilowatts. Several of these units have logged more than 40,000 hours of operational service.

In the U.S. and Japan, most PAFCs were purchased for power-generating installations that produce both heat and power. More recently, five other niche markets have sprung up, at landfills, wastewater treatment plants, food processors, power-generating facilities that cannot tolerate interruptions and green facilities (such as the aforementioned Green Building in New York City). In the first three applications, methane gas that would otherwise be an undesirable waste product is fed into the fuel cells; this free fuel helps to defray the cells' high purchase price.

The costs of PAFCs have been stalled for years now at about $4,000 per kilowatt—roughly three times what is believed to be necessary for the cells to be competitive. This fact has prompted some

observers to write off the technology as a dead end, and most fuel-cell companies established in the past three or four years are pursuing other technologies, such as molten carbonate, solid oxide and proton-exchange membrane.

Molten-carbonate fuel cells (MCFCs) and solid-oxide fuel cells (SOFCs) are similar in that they must be operated at high temperatures, in excess of 650 degrees Celsius (1,200 degrees Fahrenheit). As its name implies, the MCFC cannot operate until its electrolyte becomes molten, and the solid-oxide cells rely on the high temperatures to reform fuels internally and ionize hydrogen, without a need for expensive catalysts. On the other hand, this heat comes from the cells' output, reducing it marginally. Some engineers are envisioning residential applications in which waste heat from these cells would be captured and used to heat living spaces and water [see illustration, "Solid-Oxide Fuel Cell," page 115].

The major U.S. players in MCFCs are the Energy Research Corporation (ERC) in Danbury, Connecticut, and M-C Power Corporation in Burr Ridge, Illinois. ERC built the two-megawatt plant in Santa Clara, California, mentioned earlier. It operated for 3,000 hours; unfortunately, it rarely put out more than a megawatt. Recently, ERC changed its focus to 250-kilowatt units. M-C Power demonstrated a 250-kilowatt unit in San Diego in 1997, although it managed to produce only a disappointing 160 megawatt-hours of electricity before requiring repairs. Approximately 10 Japanese companies are also pursuing MCFCs.

As for the SOFC, a total of 40 companies around the world are developing the technology. One of the largest was created in 1998, when Siemens acquired Westinghouse Power Generation; both companies had been working on versions of the SOFC. Other important U.S. developers of the SOFC include SOFCo, ZTek Corporation and McDermott.

Proton Exchange and the New Paradigm

If the phosphoric-acid, molten-carbonate and solid-oxide fuel cells are all in some ways vestiges of the centralized paradigm of deployment, the proton-exchange membrane technology represents the burgeoning decentralized approach. There is mounting enthusiasm for PEM cells in the wake of recent, significant reductions in the cost of producing their electrolytes and of the creation of catalysts that are more resistant to the degradation caused by carbon monoxide from the reformers.

The key component of a PEM is a thin, semipermeable membrane, which functions as an electrolyte. Positively charged particles—such as hydrogen ions—can pass through this membrane, whereas electrons and atoms cannot. A few years ago developers discovered that Gore-Tex, a material often used in outer garments, can be used to strengthen the membranes and significantly improve their operating characteristics.

That and other advances have prompted a flurry of activity in the devices. At present, some 85 organizations, including 48 in the U.S., are doing research on or developing PEMs. For example, Ballard Generation Systems in Burnaby, British Columbia, is working on a modular-design PEM that can be configured up to 250 kilowatts. The company hopes to begin selling the 250-kilowatt units in 2001.

General Electric's Power Systems division and the previously mentioned Plug Power have joined together to market, install and service PEMs worldwide with capacities up to 35 kilowatts. The joint venture expects to begin field-testing prototype units later this year and to install the first residential-size units early in 2001. Plug Power installed and operates a seven-kilowatt PEM fuel cell at a home in Latham, New York (where two of the company's engineers live during the week).

Another company betting on PEMs is H Power Corporation in Belleville, New Jersey, which offers small units ranging from 35 to 500 watts. Besides promoting the cells for the usual residential uses, H Power has ventures targeting applications in backup power, telecommunications and transportation. In an unusual marketing strategy, it is even promoting fuel cells as security against the blackouts that some people fear will result from software glitches at the turn of the millennium. In another project, H Power is retrofitting 65 movable message road signs with fuel-cell power sources for the New Jersey Department of Transportation.

A few of the other PEM developers are Avista Laboratories in Spokane, Washington, which is working in conjunction with the engineering firm Black and Veatch; Matsushita Electric Industrial Company in Japan, which is focusing on a 1.5- to 3.0-kilowatt cell; and Sanyo, which has developed an appliance-like one-kilowatt PEM fuel-cell system that operates on compressed hydrogen. Sanyo also plans to develop a two-kilowatt unit using either a natural gas or methanol reformer.

Alkaline fuel cells, which have a relatively long history in exotic uses such as the space shuttle, are intriguing because they have efficiencies as high as 70 percent. So far their very high cost and other concerns have kept them out of mainstream applications. Nevertheless, a few organizations are attempting to produce alkaline units that are cost-competitive with other types of fuel cells, if not with other generation technologies.

Providing Premium Power

Other than continued subsidies, the best hope for fuel cells in the near future will be applications in which electricity is already expensive or in which waste gas can be used to fuel them. In fact, at current prices, it will probably take a combination of subsidies and unusually favorable circumstances. For instance, under an expanded federal government initiative in the U.S., purchasers of residential-size fuel-cell power plants may be eligible for federal help. In the past, federal agencies provided assistance only for the purchase of units of 100 kilowatts or more.

In the more distant future, concerns about global climate and related pressures to reduce emissions of carbon dioxide might even pave the way for large-scale use of fuel cells in the developing world. In a paper presented last year, Robert H. Williams of the Princeton University Center for Energy and Environmental Studies proposed that fuel cells might play a role in the electrification of China, whose 1.2 billion people have one of the world's lowest per capita rates of electricity usage. China has vast reserves of coal, which, Williams noted, could be turned into a supply of hydrogen-rich gas well

suited to fuel cells. The challenge would be to "decarbonize" the coal during gasification. This decarbonization would produce waste carbon dioxide—a key greenhouse gas. So engineers and geologists would have to sequester it somehow, separating it permanently from the environment.

Because of such issues, any large-scale deployment of fuel cells in a developing country could be a decade off. But in the developed world over the next several years, improvements in the proton-exchange, molten-carbonate and solid-oxide technologies will enable fuel cells to carve out new niches and expand the small ones they already occupy. As they do so, they will begin ushering in a cleaner, more environmentally benign hydrogen economy—and perhaps not a moment too soon.

—JULY 1999

The Ultimate Clean Fuel

A start-up contemplates nonpolluting cars
powered by an ingredient of soap.

JULIE WAKEFIELD

On a Saturday night, in a small garage near the Jersey shore, a mechanic turned on a Ford Explorer and put it into drive. The vehicle lunged two feet ahead. Sounds of jubilation erupted. Steven Amendola, a mustachioed chemist, jumped in the passenger seat. Five other giddy researchers piled in the back. They drove forward and backward, moving 10 feet at a time, again and again.

That joyride, possibly the world's shortest in distance, happened two years ago. Amendola's team had just proved that when dissolved in water an unassuming white powder made from borax, a common ingredient of laundry soap, could power a fuel-cell vehicle. No polluting emissions or greenhouse by-products would result from its combustion. Moreover, the basic fuel ingredient is relatively abundant.

Several months later the SUV helped to take public the fledgling company Millennium Cell, allowing it to raise $30 million. "We drove this one to the Nasdaq," quips program director Richard M. Mohring. Under the hood of the Ford Explorer is the brainchild of the Amendola-led team, the patented Hydrogen on Demand system, a compact series of pumps, tubes and catalyst chambers. When the fuel—a.k.a. sodium borohydride—contacts the catalyst, the reaction produces hydrogen gas, which, along with oxygen from air, drives most fuel cells. "We're the first to say that storing hydrogen is easy," Mohring observes. Hydrogen's volatility has been a critical hurdle to commercialization. But in the Millennium system, the hazard is reduced because the gas is produced onboard a small volume at a time and is kept at a mere fraction of the pressure of conventional compressed hydrogen supplies.

"In many regards, it's safer than gasoline," claims president and CEO Stephen Tang. The fuel itself, 7 percent hydrogen by weight, is nonflammable and nonexplosive. Another plus is that world reserves of borates, estimated at more than 600 million metric tons, could meet demand for decades, according to a recent study by U.S. Borax, a leading supplier. The known U.S. reserves rank second behind Turkey's.

The energy density of sodium borohydride—the usable energy stored in each liter of hydrogen—brings fuel cells, with their hallmark high efficiency, within range of gasoline internal-combustion engines. The latest prototypes can power a fuel-cell vehicle for 300 miles before refueling. Last December, DaimlerChrysler unveiled the Natrium, a hybrid Town & Country–concept minivan that uses a fuel-cell engine and a battery-assisted power train. In the Natrium, the Millennium technology gets its first test in a prototype vehicle, supplying hydrogen to a fuel cell made by Vancouver-based Ballard Power Systems. Now touring the country, the van ramps from zero to 60 mph in an unimpressive 16 seconds, although DaimlerChrysler is confident that better performance can be achieved.

The attractions of sodium borohydride's chemistry have been known for more than half a century. It was among the family of boranes developed as missile and jet propellants by both the U.S. and the former Soviet Union until new fuels came along in the 1950s. Today the compound is used mainly as a bleaching agent in paper mills. Working as an energy consultant in the early 1990s, Amendola grasped that sodium borohydride might be ideal for fueling the roomy gas guzzlers that Americans like to drive. A tinkerer since childhood who toyed with propellants and explosives in graduate school, he had secured two patents for cleaner coal processes before earning his Ph.D. in chemistry. With some spare sodium borohydride lying around his lab, Amendola fashioned a battery. Not only did it work on the first try, but it ran for 11 days straight. Attracting investors was easy.

To make those investments pay off, Millennium is calling on the brainpower and experience of two Nobel Prize–winning advisers and senior managers who defected from Duracell, Du Pont and Dow Chemical to bring the technology to the masses. The company must overcome a series of monumentally imposing technical barriers. Commercial sodium borohydride has a cost 50 times that of a comparable tank of gasoline. Moreover, the fuel supplies less energy than is required to produce it.

The lack of an infrastructure to provide a nonhydrocarbon energy source remains a serious stumbling block that the company alone cannot remove. Tang envisions the day when filling stations pump sodium borohydride into cars while a waste product, sodium borate, is pumped out and returned to a synthesis plant for recycling.

Millennium's fate will be tied to fuel-cell progress and broader societal factors that could eventually lead to the embrace of a nonhydrocarbon fuel. It is not just Millennium's fuel that is expensive. Savings afforded by technological advances and manufacturing economies of scale will be needed to bring the cost down for fuel cells. Yet Tang, a member of the U.S. Department of Energy's recently created Hydrogen Vision Panel, contends that forces are aligning to hasten the adoption of hydrogen technology, especially now that the Bush administration has ditched programs that advocate high-efficiency petroleum-fueled vehicles.

A series of interim technologies is needed while the company awaits the hydrogen economy. Internal-combustion engines that burn hydrogen supplied by the Millennium system instead of generating electricity with a fuel cell could be a stepping-stone to mass commercialization. Parked in the company's oversize garage is a New York City taxicab equipped with just such an engine. It emits low levels of nitrogen oxides compared with today's petroleum-burning engines and does not emit any carbon dioxide. Near-term applications for Millennium's technology may prove themselves for backup and portable power generators for a host of systems, including silicon chip factories, telecommunications networks, and "exoskeletons" (or strength-enhancing suits) for soldiers. Long-life sodium borohydride batteries and fuel-cell systems for the military could find their way into tanks, ships, unmanned air vehicles and more. Supply-chain advantages abound: marine vehicles, for example, could draw water from the ocean and

mix it onboard with dry fuel to enable longer stints at sea before refueling.

Clean-car concepts come in many varieties, of course, including natural gas and methanol. A number of researchers predict that the ultimate answer will be a solid-state medium for hydrogen storage, perhaps using carbon nanotubes, a development still decades away. Whether Millennium emerges as more than a niche player hinges on its ability to line up the right partners early, according to David Sackler, vice president of Vestigo Associates, the research arm of Fidelity Capital Markets. So far the company has inked deals with U.S. Borax, Rohm and Haas, and Air Products and Chemicals on the supply side and is attracting various auto manufacturers on the demand front. Signing on a big energy company or automaker to build a large-scale production facility will be critical, Sackler says.

If Millennium can form the right alliances and bring down energy costs for more than a single Natrium car, its fuel technology could help usher in the new era, Tang predicts. If he gets his way, the white powder will power everything from small batteries for personal computers and wheelchairs to engines and large fuel cells in buses, ferries, submarines and perhaps even aircraft. All this musing brings a fresh buzz to a chemical whose prior fame stemmed from its association with Ronald Reagan's hawking of 20 Mule Team Borax cleanser on the television program *Death Valley Days*.

—MAY 2002

Questions about a Hydrogen Economy

Much excitement surrounds the progress in fuel cells, but the quest for a hydrogen economy is no trivial pursuit.

MATTHEW L. WALD

In the fall of 2003, a few months after President George W. Bush announced a $1.7 billion research program to develop a vehicle that would make the air cleaner and the country less dependent on

Overview: Hydrogen Economy

- Per a given equivalent unit of fuel, hydrogen fuel cells in vehicles are about twice as efficient as internal-combustion engines. Unlike conventional engines, fuel cells emit only water vapor and heat.

- Hydrogen doesn't exist freely in nature, however, so producing it depends on current energy sources. Sources of hydrogen are either expensive and not widely available (including electrolysis using renewables such as solar, wind or hydropower), or else they produce undesirable greenhouse gases (coal or other fossil fuels).

- Ultimately hydrogen may not be the universal cure-all, although it may be appropriate for certain applications. Transportation may not be one of them.

imported oil, Toyota came to Washington, D.C., with two of them. One, a commercially available hybrid sedan, had a conventional, gasoline-fueled internal-combustion engine supplemented by a battery-powered electric motor. It got about 50 miles to the gallon, and its carbon dioxide emissions were just over half those of an average car. The other auto, an experimental SUV, drove its electric motor with hydrogen fuel cells and emitted as waste only water purer than Perrier and some heat. Which was cleaner?

Answering that question correctly could have a big impact on research spending, on what vehicles the government decides to subsidize as it tries to incubate a technology that will wean us away from gasoline and, ultimately, on the environment. But the answer is not what many people would expect, at least according to Robert Wimmer, research manager for technical and regulatory affairs at Toyota. He said that the two vehicles were about the same.

Wimmer and an increasing number of other experts are looking beyond simple vehicle emissions, to the total effect on the environment caused by the production of the vehicle's fuel and its operation combined. Seen in a broader context, even the supposed great advantages of hydrogen, such as the efficiency and cleanliness of fuel cells, are not as overwhelming as might be thought. From this perspective, coming in neck and neck with a hybrid is something of an achievement; in some cases, the fuel-cell car can be responsible for substantially more carbon dioxide emissions, as well as a variety of other pollutants, the Department of Energy states. And in one way the hybrid is, arguably, superior: it already exists as a commercial product and thus is available to cut pollution now. Fuel-cell cars, in contrast, are expected on about the same schedule as NASA's manned trip to Mars and have about the same level of likelihood.

If that sounds surprising, it is also revealing about the uncertainties and challenges that trail the quest for a hydrogen economy—wherein most energy is devoted to the creation of hydrogen, which is then run through a fuel cell to make electricity. Much hope surrounds the advances in fuel cells and the possibility of a cleaner hydrogen economy, which could include not only transportation but also power for houses and other buildings. Last November U.S. Energy Secretary Spencer Abraham told a Washington gathering of energy ministers from 14 countries and the European Union that hydrogen could "revolutionize the world in which we live." Noting that the nation's more than 200 million motor vehicles consume about two-thirds of the 20 million barrels of oil the U.S. uses every day, President Bush has called hydrogen the "freedom fuel."

But hydrogen is not free, in either dollars or environmental damage. The hydrogen fuel cell costs nearly 100 times as much per unit of power produced as an internal-combustion engine. To be price competitive, "you've got to be at a nickel a watt, and we're at $4 a watt," says Tim R. Dawsey, a research associate at Eastman Chemical Company, which makes polymers for fuel cells. Hydrogen is also about five times as expensive, per unit of usable energy, as gasoline. Simple dollars are only one speed bump on the road to the hydrogen economy. Another is that supplying the energy required to make pure hydrogen may itself cause pollution. Even if that energy is from a renewable source, like the sun or the wind, it may have more environmentally sound uses than the production of hydrogen. Distribution and storage of hydrogen—the least dense gas in the universe—are other technological and infrastructure difficulties. So is the safe handling of the gas. Any practical proposal for a hydrogen economy will have to address all these issues.

Which Sources Make Sense?

Hydrogen fuel cells have two obvious attractions. First, they produce no pollution at point of use [see "Vehicle of Change," page 153]. Second, hydrogen can come from myriad sources. In fact, the gas is not a fuel in the conventional sense. A fuel is something found in nature, like coal, or refined from a natural product, like diesel fuel from oil, and then burned to do work. Pure hydrogen does not exist naturally on earth and is so highly processed that it is really more of a carrier or medium for storing and transporting energy from some original source to a machine that makes electricity. "The beauty of hydrogen is the fuel diversity that's possible," said David K. Garman, U.S. assistant secretary for energy efficiency and renewable energy. Each source, however, has an ugly side.

For instance, a process called electrolysis makes hydrogen by splitting a water molecule with electricity [see sidebar, "Creating Hydrogen," page 127]. The electricity could come from solar cells, windmills, hydropower or safer, next-generation nuclear reactors [see "Next-Generation Nuclear Power," page 75]. Researchers are also trying to use microbes to transform biomass, including parts of crops that now have no economic value, into hydrogen. In February researchers at the University

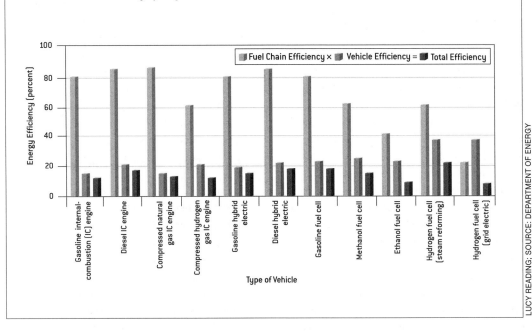

Well-to-Wheels Energy Efficiency

Total energy efficiency includes not only vehicle operation but also the energy required to produce fuel. Extracting oil, refining gasoline and trucking that fuel to filling stations for internal-combustion engines is more efficient than creating hydrogen for fuel cells.

Energy Efficiency (percent)

Legend: ■ Fuel Chain Efficiency × ■ Vehicle Efficiency = ■ Total Efficiency

Type of Vehicle

Vehicle types (left to right): Gasoline internal-combustion (IC) engine; Diesel IC engine; Compressed natural gas IC engine; Compressed hydrogen gas IC engine; Gasoline hybrid electric; Diesel hybrid electric; Gasoline fuel cell; Methanol fuel cell; Ethanol fuel cell; Hydrogen fuel cell (steam reforming); Hydrogen fuel cell (grid electric)

LUCY READING; SOURCE: DEPARTMENT OF ENERGY

of Minnesota and the University of Patras in Greece announced a chemical reactor that generates hydrogen from ethanol mixed with water. Though appealing, all these technologies are either unaffordable or unavailable on a commercial scale and are likely to remain so for many years to come, according to experts.

Hydrogen could be derived from coal-fired electricity, which is the cheapest source of energy in most parts of the country. Critics argue, though, that if coal is the first ingredient for the hydrogen economy, global warming could be exacerbated through greater release of carbon dioxide.

Or hydrogen could come from the methane in natural gas, methanol or other hydrocarbon fuel [see sidebar, "Creating Hydrogen," page 127]. Natural gas can be reacted with steam to make

hydrogen and carbon dioxide. Filling fuel cells, however, would preclude the use of natural gas for its best industrial purpose today: burning in high-efficiency combined-cycle turbines to generate electricity. That, in turn, might again lead to more coal use. Combined-cycle plants can turn 60 percent of the heat of burning natural gas into electricity; a coal plant converts only about 33 percent. Also, when burned, natural gas produces just over half as much carbon dioxide per unit of heat as coal does, 117 pounds per million Btu versus 212. As a result, a kilowatt-hour of electricity made from a new natural gas plant has slightly over one-fourth as much carbon dioxide as a kilowatt-hour from coal. (Gasoline comes between coal and natural gas, at 157 pounds of carbon dioxide per million Btu.) In sum, it seems better for the

Creating Hydrogen

Two main methods are known for extracting hydrogen, which does not occur in pure form naturally on the earth. Electrolysis (left) uses electric current to split molecules of water (H_2O). A cathode (negative terminal) attracts hydrogen atoms, and an anode (positive) attracts oxygen; the two gases bubble up into air and can be captured. In steam reforming (right), a hydrocarbon such as methanol (CH_3OH) first vaporizes in a heated combustion chamber. A catalyst in the steam reformer breaks apart fuel and water vapor to produce components including hydrogen, which is then separated and routed to a fuel cell.

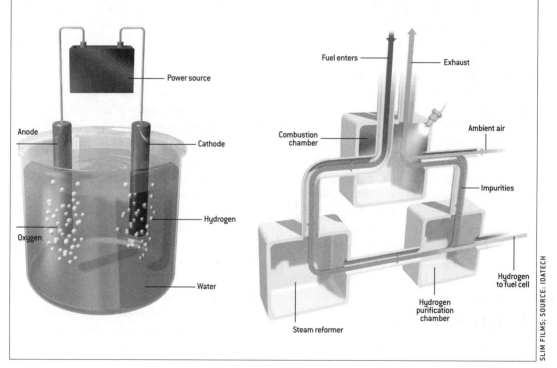

ELECTROLYSIS

- Power source
- Anode
- Cathode
- Hydrogen
- Oxygen
- Water

STEAM REFORMING

- Fuel enters
- Exhaust
- Combustion chamber
- Ambient air
- Impurities
- Hydrogen to fuel cell
- Hydrogen purification chamber
- Steam reformer

SLIM FILMS; SOURCE: IDATECH

environment to use natural gas to make electricity for the grid and save coal, rather than turning it into hydrogen to save gasoline.

Two other fuels could be steam reformed to give off hydrogen: the oil shipped from Venezuela or the Persian Gulf and, again, the coal from Appalachian mines. To make hydrogen from fossil fuels in a way that does not add to the release of climate-changing carbon dioxide, the carbon must be captured so that it does not enter the atmosphere. Presumably this process would be easier than sequestering carbon from millions of tailpipes. Otherwise, the fuels might as well be burned directly.

"If you look at it from the whole system, not the individual sector, you may do better to get rid of your coal-fired power plants, because coal is such a carbon-intensive fuel," says Michael Wang, an energy researcher at Argonne National Laboratory. Coal accounts for a little more than half the kilowatt-hours produced in the U.S.; about 20 percent is from natural gas. The rest comes from mostly carbon-free sources, primarily nuclear reactors and hydroelectricity. Thus, an effort to replace the coal-fired electric plants would most likely take decades.

In any case, if hydrogen were to increase suddenly in supply, fuel cells might not even be the best use for the gas. In a recent paper, Reuel Shinnar, professor of chemical engineering at the City College of New York, reviewed the alternatives for power and fuel production. Rather than the use of hydrogen as fuel, he suggested something far simpler: increased use of hydrocracking and hydrotreating. The U.S. could save three million barrels of oil a day that way, Shinnar calculated. Hydrocracking and hydrotreating both start with molecules in crude oil that are unsuitable for gasoline because they are too big and have a carbon-to-hydrogen ratio that is too heavy with carbon. The processes are expensive but still profitable, because they allow the refineries to take ingredients that are good for only low-value products, such as asphalt and boiler fuel, and turn them into gasoline. It is like turning chuck steak into sirloin.

What about Conversion Costs?

If hydrogen production is dirty and expensive, could its impressive energy efficiency at point of use make up for those downsides? Again, the answer is complicated.

A kilo of hydrogen contains about the same energy as a gallon of unleaded regular gas—that is, if burned, each would give off about the same amount of heat. But the internal-combustion engine and the fuel cell differ in their ability to extract usable work from that fuel energy. In the

engine, most of the energy flows out of the tailpipe as heat, and additional energy is lost to friction inside the engine. In round numbers, advocates and detractors agree, a fuel cell gets twice as much work out of a kilo of hydrogen as an engine gets out of a gallon of gas. (In a stationary application—such as a basement appliance that takes the hydrogen from natural gas and turns it into electricity to run the household—efficiency could be higher, because the heat given off by the fuel-cell process could also be used; for example, to heat tap water.)

There is, in fact, a systematic way to evaluate where best to use each fuel. A new genre of energy analysis, "well to wheels," compares the energy efficiency of every known method to turn a vehicle's wheels [see sidebar, "Well-to-Wheels Energy Efficiency," page 126]. The building block of the well-to-wheels performance is "conversion efficiency." At every step of the energy chain, from pumping oil out of the ground to refining it to burning it in an engine, some of the original energy potential of the fuel is lost.

The first part of the well-to-wheels determination is what engineers call "well to tank": what it takes to make and deliver a fuel. When natural gas is cracked for hydrogen, about 40 percent of the original energy potential is lost in the transfer, according to the DOE Office of Energy Efficiency and Renewable Energy. Using electricity from the grid to make hydrogen by electrolysis of water causes a loss of 78 percent. (Despite the lower efficiency of electrolysis, it is likely to predominate in the early stages of a hydrogen economy because it is convenient—producing the hydrogen where it is needed and thus avoiding shipping problems.) In contrast, pumping a gallon of oil out of the ground, taking it to a refinery, turning it into gasoline and getting that petrol to a filling station loses about 21 percent of the energy potential. Producing natural gas and compressing it in a tank loses only about 15 percent.

The second part of the total energy analysis is "tank to wheels," or the fraction of the energy value in the vehicle's tank that actually ends up driving the wheels. For the conventional gasoline internal-combustion engine, 85 percent of the energy in the gasoline tank is lost; thus, the whole system, well to tank combined with tank to wheels, accounts for a total loss of 88 percent.

The fuel cell converts about 37 percent of the hydrogen's energy value to power for the wheels. The total loss, well to wheels, is about 78 percent if the hydrogen comes from steam-reformed natural gas. If the source of the hydrogen is electrolysis from coal, the loss from the well (a mine, actually) to tank is 78 percent; after that hydrogen runs through a fuel cell, it loses another 43 percent, with the total loss reaching 92 percent.

Wally Rippel, a research engineer at AeroVironment in Monrovia, California, who helped to develop the General Motors EV-1 electric car and the NASA Helios Solar Electric airplane, offers another way to look at the situation. He calculates that in a car that employs an electric motor to turn the wheels, a kilowatt-hour used to recharge batteries will propel the auto three times as far as if that same kilowatt-hour were instead used to make hydrogen for a fuel cell.

All these facts add up to an argument *not* to use electricity to make hydrogen and then go back to electricity again with an under-the-hood fuel cell. But there is one strong reason to go through inefficient multiple conversions. They may still make economic sense, and money is what has shaped the energy markets so far. That is, even if the hydrogen system is very wasteful of energy, there are such huge differences in the cost of energy from various sources that it might make sense to switch to a system that lets us go where the cheapest energy is.

Walter "Chip" Schroeder, president and chief executive of Proton Energy Systems, a Connecticut company that builds electrolysis machines, explains the economic logic. Coal at current prices (which is to say, coal at prices that are likely to prevail for years to come) costs a little more than 80 cents per million Btu. Gasoline at $1.75 a gallon (which seems pricey at the moment but in a few months or years could look cheap) is about $15.40. The mechanism for turning a Btu from coal into a Btu that will run a car is cumbersome, but in the transition, "you end up with wine, not water," he says. Likewise, he describes his device to turn water into hydrogen as an "arbitrage machine." "Arbitrage" is the term used by investment bankers or stock or commodities traders to describe buying low and selling high, but it usually refers to small differences in the price of a stock or the value of a currency between one market or another. "You can't make reasonable policy without understanding just how extreme the value differentials in our energy marketplace are," Schroeder says.

How to Deliver the Hydrogen?

Different sources of energy may not be as fungible as money is in arbitrage, however. There is a problem making hydrogen conveniently available at a good cost, at least if the hydrogen is going to come from renewable sources such as solar, hydropower or wind that are practical in only certain areas of the country.

Hydrogen from wind, for example, is competitive with gasoline when wind power costs three cents a kilowatt-hour, says Garman of the DOE. That occurs where winds blow steadily. "Where I might get three-cent wind tends to be in places where people don't live," he notes. In the U.S., such winds exist in a belt running from Montana and the Dakotas to Texas. The electric power they produce would have a long way to go to reach the end users—with energy losses throughout the grid along the way. "You can't get the electrons out of the Dakotas because of transmission constraints," Garman points out. "Maybe a hydrogen pipeline could get the tremendous wind resource carried to Chicago," the nearest motor-fuel market.

That is, if such a pipeline were even practical to build. Given hydrogen's low density, it is far harder to deliver than, for instance, natural gas. To move large volumes of any gas requires compressing it, or else the pipeline has to have a diameter similar to that of an airplane fuselage. Compression takes work, and that drains still more energy from the total production process. Even in this instance, managing hydrogen is trickier than dealing with other fuel gases. Hydrogen compressed to about 790 atmospheres has less than a third of the energy of the methane in natural gas at the same pressure, points out a recent study by three European researchers, Ulf Bossel, Baldur Eliasson and Gordon Taylor.

A related problem is that a truck that could deliver 2,400 kilos of natural gas to a user would yield only 288 kilos of hydrogen pressurized to the same level, Bossel and his colleagues find. Put another way, it would take about 15 trucks to deliver the hydrogen needed to power the same number of cars that could be served by a single gasoline tanker. Switch to liquid hydrogen, and it would take only about three trucks to equal the one gasoline tanker, but hydrogen requires substantially more effort to liquefy. Shipping the hydrogen as methanol that could be reformed onboard the vehicle [see sidebar, "Creating Hydrogen," page 127] would ease transport, but again, the added transition has an energy penalty. These facts argue for using the hydrogen where it is produced, which may be distant from the major motor-fuel markets.

No matter how hydrogen reaches its destination, the difficulties of handling the elusive gas will not be over. Among hydrogen's disadvantages is that it burns readily. All gaseous fuels have a minimum and maximum concentration at which they will burn. Hydrogen's range is unusually broad, from 2 to 75 percent. Natural gas, in contrast,

burns between 5 and 15 percent. Thus, as dangerous as a leak of natural gas is, a hydrogen leak is worse, because hydrogen will ignite at a wider range of concentrations. The minimum energy necessary to ignite hydrogen is also far smaller than that for natural gas.

And when hydrogen burns, it does so invisibly. NASA published a safety manual that recommends checking for hydrogen fires by holding a broom at arm's length and seeing if the straw ignites. "It's scary—you cannot see the flame," says Michael D. Amiridis, chair of the department of chemical engineering at the University of South Carolina, which performs fuel-cell research under contract for a variety of companies. A successful fuel-cell car, he says, would have "safety standards at least equivalent to the one I have now." A major part of the early work on developing a hydrogen fueling supply chain has been building warning instruments that can reliably detect hydrogen gas.

A Role for Hydrogen

Despite the technological and infrastructure obstacles, a hydrogen economy may be coming. If it is, it will most likely resemble the perfume economy, a market where quantities are so small that unit prices do not matter. Chances are good that it will start in cellular phones and laptop computers, where consumers might not mind paying $10 a kilowatt-hour for electricity from fuel cells; a recent study by the fuel-cell industry predicts that the devices could be sold in laptop computers this year. It might eventually move to houses, which will run nicely on five kilowatts or so and where an improvement in carbon efficiency is highly desirable because significant electricity demand exists almost every hour of the day. But hydrogen cells may not appear in great numbers in driveways, where cars have a total energy requirement of about 50 kilowatts apiece but may run only an average of two hours a day—a situation that is exactly backward from where a good engineer would put a device like a fuel cell, which has a low operating cost but a high cost per unit of capacity. Although most people may have heard of fuel cells as alternative power sources for cars, cars may be the last place they'll end up on a commercial scale.

If we need to find substitutes for oil for transportation, we may look to several places before hydrogen. One is natural gas, with very few technical details to work out and significant supplies available. Another is electricity for electric cars. Battery technology has hit some very significant hurdles, but they might be easier to solve than those of fuel cells. If we have to, we can run vehicles on methanol from coal; the Germans did it in the 1940s, and surely we could figure it out today.

Last, if we as a society truly support the development of renewable sources such as windmills and solar cells, they could replace much of the fossil fuels used today in the electric grid system. With that development, plus judicious conservation, we would have a lot of energy left over for the transportation sector, the part of the economy that is using up the oil and making us worry about hydrogen in the first place.

—MAY 2004

A Power Grid for the Hydrogen Economy

Cryogenic, superconducting conduits could be connected into a "SuperGrid" that would simultaneously deliver electrical power and hydrogen fuel.

PAUL M. GRANT, CHAUNCEY STARR
AND THOMAS J. OVERBYE

On the afternoon of August 14, 2003, electricity failed to arrive in New York City, plunging the eight million inhabitants of the Big Apple—along with 40 million other people throughout the northeastern U.S. and Ontario—into a tense night of darkness. After one power plant in Ohio had shut down, elevated power loads overheated high-voltage lines, which sagged into trees and short-circuited. Like toppling dominoes, the failures cascaded through the electrical grid, knocking 265 power plants offline and darkening 24,000 square kilometers.

That incident—and an even more extensive blackout that affected 56 million people in Italy and Switzerland a month later—called attention to pervasive problems with modern civilization's vital equivalent of a biological circulatory system, its interconnected electrical networks. In North America the electrical grid has evolved in piecemeal fashion over the past 100 years. Today the more

Overview: A Continental SuperGrid

- As the 2003 blackouts in North America and Europe vividly testify, the current power grid is struggling to meet growing demand for electricity and the coming shift from fossil-fueled power and cars to cleaner sources of energy.

- For several years, engineers have been designing a new infrastructure that would enable cities to tap power efficiently from large nuclear and renewable energy plants in distant and remote locations.

- SuperCables would transmit extraordinarily high electrical current nearly resistance-free through superconducting wires. The conduits would also carry ultracold hydrogen as a liquid or high-pressure gas to factories, vehicle fueling stations and perhaps one day even to home furnaces and boilers.

SuperCables

SuperCables could transport energy in both electrical and chemical form. Electricity would travel nearly resistance-free through pipes made of a superconducting material. Chilled hydrogen flowing as a liquid inside the conductors would keep their temperature near absolute zero. A SuperCable with two conduits, each about a meter in diameter, could simultaneously transmit five gigawatts of electricity and 10 gigawatts of thermal power *(table)*.

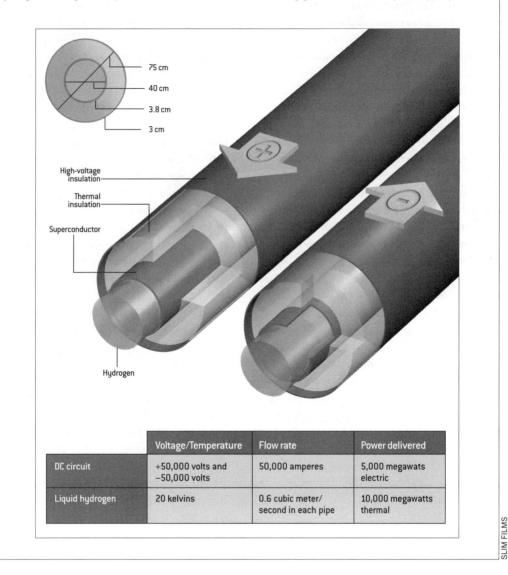

	Voltage/Temperature	Flow rate	Power delivered
DC circuit	+50,000 volts and −50,000 volts	50,000 amperes	5,000 megawats electric
Liquid hydrogen	20 kelvins	0.6 cubic meter/ second in each pipe	10,000 megawatts thermal

SLIM FILMS

than $1 trillion infrastructure spans the continent with millions of kilometers of wire operating at up to 765,000 volts. Despite its importance, no single organization has control over the operation, maintenance or protection of the grid; the same is true in Europe. Dozens of utilities must cooperate even as they compete to generate and deliver, every second, exactly as much power as customers demand—and no more. The 2003 blackouts raised calls for greater government oversight and spurred the industry to move more quickly, through its IntelliGrid Consortium and the GridWise program of the U.S. Department of Energy, to create self-healing systems for the grid that may prevent some kinds of outages from cascading. But reliability is not the only challenge—and arguably not even the most important challenge—that the grid faces in the decades ahead.

A more fundamental limitation of the 20th-century grid is that it is poorly suited to handle two 21st-century trends: the relentless growth in demand for electrical energy and the coming transition from fossil-fueled power stations and vehicles to cleaner sources of electricity and transportation fuels. Utilities cannot simply pump more power through existing high-voltage lines by ramping up the voltages and currents. At about one million volts, the electric fields tear insulation off the wires, causing arcs and short circuits. And higher currents will heat the lines, which could then sag dangerously close to trees and structures.

It is not at all clear, moreover, how well today's infrastructure could support the rapid adoption of hybrid vehicles that draw on electricity or hydrogen for part of their power. And because the power system must continuously match electricity consumption with generation, it cannot easily accept a large increase in the unpredictable and intermittent power produced from renewable wind, ocean and solar resources.

We are part of a growing group of engineers and physicists who have begun developing designs for a new energy delivery system we call the Continental SuperGrid. We envision the SuperGrid evolving gradually alongside the current grid, strengthening its capacity and reliability. Over the course of decades, the SuperGrid would put in place the means to generate and deliver not only plentiful, reliable, inexpensive and "clean" electricity but also hydrogen for energy storage and personal transportation.

Engineering studies of the design have concluded that no further fundamental scientific discoveries are needed to realize this vision. Existing nuclear, hydrogen and superconducting technologies, supplemented by selected renewable energy, provide all the technical ingredients required to create a SuperGrid. Mustering the social and national resolve to create it may be a challenge, as will be some of the engineering. But the benefits would be considerable, too.

Superconducting lines, which transmit electricity with almost perfect efficiency, would allow distant generators to compensate for local outages. They would allow power plants in different climate regions to bolster those struggling to meet peak demand. And they would allow utilities to construct new generating stations on less controversial sites far from population centers.

SuperGrid connections to these new power plants would provide both a source of hydrogen and a way to distribute it widely, through pipes that surround and cool the superconducting wires. A hydrogen-filled SuperGrid would serve not only as a conduit but also as a vast repository of energy, establishing the buffer needed to enable much more extensive use of wind, solar and other renewable power sources. And it would build the core infrastructure that is a prerequisite if rich economies are to move away from greenhouse gas–emitting power plants and vehicles.

A New Grid for a New Era

A continental supergrid may sound like a futuristic idea, but the concept has a long history. In 1967 IBM physicists Richard L. Garwin and Juri Matisoo published a design for a 1,000-kilometer transmission cable made of niobium tin, which superconducts at high currents. Extraordinary amounts of direct current (DC) can pass resistance-free through such a superconductor when the metal is chilled by liquid helium to a few degrees above absolute zero. The scientists proposed a DC cable with two conductors (made of superconducting wire or tape) that together would carry 100 gigawatts—roughly the output of 50 nuclear power plants.

Garwin and Matisoo were exploring what might be possible, not what would be practical. It would not make sense to inject that much power into one point of the grid, and liquid helium is a cumbersome coolant. But their ideas inspired others. In the following decades, short superconducting cables were built and tested to carry alternating current (AC) in Brookhaven, New York, and near Graz, Austria, with the latter operating connected to the local grid for several years.

Ten years after the discovery of high-temperature superconductivity, a technical study by the Electric Power Research Institute (EPRI) concluded that with liquid nitrogen as a coolant, a five-gigawatt DC "electricity pipe" could compete economically with a gas pipeline or conventional overhead lines for transmission distances of 800 kilometers or more. Two of us (Grant and Starr) developed the idea further in papers that explored how ultracold hydrogen—either liquid or supercritical gas—might both chill the superconducting wires and deliver energy in chemical form within a continental-scale system. In 2002 and 2004 the third author (Overbye) organized workshops at which dozens of experts detailed a plan for a 100-meter pilot seg-

ment, precursor to a 50-kilometer intertie between existing regional grids.

It is important to develop prototypes soon, because existing electrical grids are increasingly reaching the point of maximum loading—and, as the blackouts indicate, occasionally exceeding it. As total generating capacity in the U.S. has risen by almost a quarter in the past five years, the high-voltage transmission grid has grown in size by just 3.3 percent. Yet society's appetite for energy continues to grow rapidly: the U.S. Energy Information Administration forecasts that by 2025 annual energy use in the U.S. will hit 134 trillion mega-joules (127 quadrillion Btus), over a quarter greater than it was in 2005.

The rising demand poses two problems: where to get this new energy and how to distribute it. Fossil fuels will probably still supply a large fraction of our energy 20 years from now. But global competition for limited petroleum and natural gas resources is intense, and even mild production shortages can send prices skyrocketing, as we have seen in the past few months. Concern over greenhouse warming is leading to other constraints.

If we have an opportunity to move away from our dependence on fossil fuels, clearly we should take it. But fully exploiting nonfossil energy sources, including wind, solar, agricultural biomass and in particular advanced nuclear power, will require a new grid for this new era. To distribute trillions of kilowatt-hours of extra electricity every year, the U.S. grid will have to handle roughly 400 gigawatts more power than it does today.

The current infrastructure can be enhanced only so far. New carbon-core aluminum wires can be stretched more tautly than conventional copper wires and so can carry perhaps three times as much current before sagging below safe heights. And U.S. utilities will take advantage of provisions in the 2005 Energy Act that make it easier to open new transmission corridors.

The Evolution of a SuperGrid

Transition to a SuperGrid would take at least a generation to complete. The evolution would inject new technologies into every level of the infrastructure: generators, transformers, power transmission and consumption.

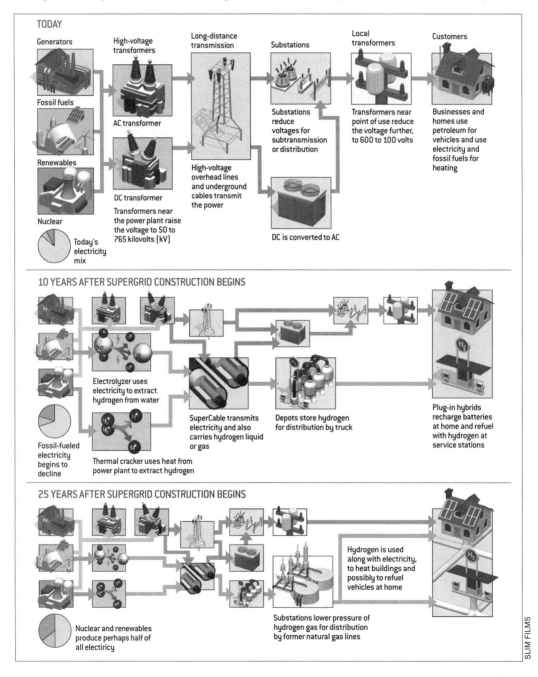

TODAY

Generators

Fossil fuels

Renewables

Nuclear

Today's electricity mix

High-voltage transformers

AC transformer

DC transformer

Transformers near the power plant raise the voltage to 50 to 765 kilovolts (kV)

Long-distance transmission

High-voltage overhead lines and underground cables transmit the power

Substations

Substations reduce voltages for subtransmission or distribution

DC is converted to AC

Local transformers

Transformers near point of use reduce the voltage further, to 600 to 100 volts

Customers

Businesses and homes use petroleum for vehicles and use electricity and fossil fuels for heating

10 YEARS AFTER SUPERGRID CONSTRUCTION BEGINS

Electrolyzer uses electricity to extract hydrogen from water

Fossil-fueled electricity begins to decline

Thermal cracker uses heat from power plant to extract hydrogen

SuperCable transmits electricity and also carries hydrogen liquid or gas

Depots store hydrogen for distribution by truck

Plug-in hybrids recharge batteries at home and refuel with hydrogen at service stations

25 YEARS AFTER SUPERGRID CONSTRUCTION BEGINS

Nuclear and renewables produce perhaps half of all electiricy

Substations lower pressure of hydrogen gas for distribution by former natural gas lines

Hydrogen is used along with electricity, to heat buildings and possibly to refuel vehicles at home

SLIM FILMS

But high-voltage lines are already approaching the million-volt limit on insulators and the operating limits of semiconductor devices that control DC lines. AC lines become inefficient at distances around 1,200 kilometers, because they begin to radiate the 60-hertz power they carry like a giant antenna. Engineers will thus need to augment the transmission system with new technologies to transport hundreds more gigawatts from remote generators to major cities.

Next-Generation Nuclear

One of our goals in designing the SuperGrid has been to ensure that it can accept inputs from a wide variety of generators, from the smallest rooftop solar panel and farmyard wind turbine to the largest assemblage of nuclear reactors. The largest facilities constrain many basic design decisions, however. And the renewables still face tremendous challenges in offering the enormous additional capacity required for the next 20 years. So we built our concept on a foundation of fourth-generation nuclear power.

The 2005 Energy Act directed $60 million toward development of "generation IV" high-temperature, gas-cooled reactors. Unlike most current nuclear plants, which are water-cooled and so usually built near large bodies of water—typically near population centers—the next-generation reactors expel their excess heat directly into the air or earth.

In newer designs, the nuclear reactions slow down as the temperature rises above a normal operating range. They are thus inherently resistant to the coolant loss and overheating that occurred at Chernobyl in Ukraine and Three Mile Island in Pennsylvania [see "Next-Generation Nuclear Power," page 75].

Like all fission generators, however, Generation IV units will produce some radioactive waste. So it will be least expensive and easiest politically to build them in "nuclear clusters," far from urban areas. Each cluster could produce on the order of 10 gigawatts.

Remote siting will make it easier to secure the reactors as well as to build them. But we will need a new transmission technology—a SuperCable— that can drastically reduce the cost of moving energy over long distances.

SuperCable

For the electricity part of the SuperGrid, where we need to move tens of gigawatts over hundreds of kilometers, perfect conductors are a perfect fit. Although superconducting materials were discovered in 1911 and were fashioned into experimental devices decades ago, it is only quite recently that the refrigeration needed to keep them ultracold has become simple enough for industrial use. Superconductors are now moving beyond magnetic resonance imaging scanners and particle accelerators and into commercial power systems.

For example, the DOE has joined with power equipment manufacturers and utilities to produce prototypes of superconducting transformers, motors, generators, fault-current limiters and transmission cables. Other governments—notably Japan, the European Union, China and South Korea—have similar development programs. Three pilot projects now under way in the U.S. are demonstrating superconducting cables in Columbus, Ohio, and in New York State on Long Island and in Albany.

These cables use copper oxide–based superconducting tape cooled by liquid nitrogen at 77 kelvins (–196 degrees Celsius). Using liquid hydrogen for coolant would drop the temperature to 20 kelvins, into the superconducting range of new compounds such as magnesium diboride [see "Low-Temperature Superconductivity Is Warming Up," by Paul C. Canfield and Sergey L. Bud'ko; *Scientific American*, April 2005].

All demonstrations of superconducting cables so far have used AC power, even though only DC electricity can travel without resistance. Even so, at the frequencies used on the current grid, superconductors offer about one two-hundredth the electrical resistance of copper at the same temperature.

The SuperCable we have designed includes a pair of DC superconducting wires, one at plus 50,000 volts, the other at minus 50,000 volts, and both carrying 50,000 amps—a current far higher than any conventional wire could sustain. Such a cable could transmit about five gigawatts for several hundred kilometers at nearly zero resistance and line loss. (Today about a tenth of all electrical energy produced by power plants is lost during transmission.)

A five-gigawatt SuperCable is certainly technically feasible. Its scale would rival the 3.1-gigawatt Pacific Intertie, an existing 500-kilovolt DC overhead line that moves power between northern Oregon and southern California. Just four SuperCables would provide sufficient capacity to transmit all the power generated by the giant Three Gorges Dam hydroelectric facility in China.

Because a SuperCable would use hydrogen as its cryogenic coolant, it would transport energy in chemical as well as electrical form. Next-generation nuclear plants can produce either electricity or hydrogen with almost equal thermal efficiency. So the operators of nuclear clusters could continually adjust the proportions of electricity and "hydricity" that they pump into the SuperGrid to keep up with the electricity demand while maintaining a flow of hydrogen sufficient to keep the wires superconducting.

Electricity and Hydricity

The ability to choose among alternative forms of power and to store electricity in chemical form opens up a world of possibilities. The SuperGrid could dramatically reduce fuel costs for electric- and hydrogen-powered hybrid vehicles, for example.

Existing hybrids run on gasoline or diesel but use batteries to recover energy that otherwise would go to waste. "Plug-in" hybrids that debuted last year use electricity as well as gas [see "Hybrid Vehicles," by Joseph J. Romm and Andrew A. Frank; *Scientific American,* April 2006]. BMW, Mazda and others have demonstrated hydrogen hybrids that have two fuel tanks and engines that burn hydrogen when it is available and gasoline when it is not. Many automakers are also developing vehicles that use onboard fuel cells to turn hydrogen back into electricity by combining it with oxygen.

Even the most efficient automobiles today convert only 30 to 35 percent of their fuel energy into motion. Hydrogen fuel-cell hybrids could do significantly better, reaching 50 percent efficiencies with relative ease and eventually achieving 60 to 65 percent fuel efficiencies.

Replacing even a modest percentage of petroleum-based transportation fuels would require enormous amounts of both hydrogen and electricity, as well as a pervasive and efficient delivery infrastructure. The SuperGrid offers one way to realize this vision. Within each nuclear cluster, some reactors could produce electricity while others made hydrogen—without emitting any greenhouse gases.

By transporting the two together, the grid would serve both as a pipeline and as an energy store. For example, every 70-kilometer section of SuperCable containing 40-centimeter-diameter pipes filled with liquid hydrogen would store 32 gigawatt-hours of energy. That is equivalent to the capacity of the Raccoon Mountain reservoir, the largest pumped hydroelectric facility in the U.S.

By transforming electricity into a less ephemeral commodity similar to oil or natural gas, the new grid could allow electricity markets to tolerate rapid swings in demand more reliably than they do today. SuperGrid links crossing several time zones and weather boundaries would allow power plants to tap excess nighttime capacity to meet the peak electricity needs of distant cities. By smoothing out fluctuations in demand, the low-loss grid could help reduce the need for new generation construction.

The SuperGrid could go a long way, too, toward removing one of the fundamental limitations to the large-scale use of inconstant energy from wind, tides, waves and sunlight. Renewable power plants could pump hydrogen onto the grid, rather than selling electricity. Alternatively, baseline generators could monitor the rise and fall in electrical output from these plants and might be able to use electrolysis to shift their electricity/hydricity blend to compensate.

Charging Ahead

No major scientific advances are needed to begin building the SuperGrid, and the electric utility industry has already shown its interest in the concept by funding a SuperGrid project at EPRI that will explore the numerous engineering challenges that integrating SuperCables into the existing power grid will pose. The largest of these is what to do if a SuperCable fails.

The grid today remains secure even when a single device, such as a high-voltage transmission line, fails. When a line sags into a tree, for example, circuit breakers open to isolate the line from the grid, and the power that was flowing on the wire almost instantaneously shifts to other lines. But we do not yet have a circuit-breaker design that can cut off the extraordinary current that would flow over a SuperCable. That technology will have to evolve. Grid managers may need to develop novel techniques for dealing with the substantial disturbance that loss of such a huge amount of power would cause on the conventional grid. A break in a SuperCable would collapse the surrounding magnetic field, creating a brief but intense voltage

spike at the cut point. The cables will need insulation strong enough to contain this spike.

Safely transporting large amounts of hydrogen within the SuperCable poses another challenge. The petrochemical industry and space programs have extensive experience pumping hydrogen, both gaseous and liquid, over kilometer-scale pipelines. The increasing use of liquefied natural gas will reinforce that technology base further. The explosive potential (energy content per unit mass) of hydrogen is about twice that of the methane in natural gas. But hydrogen leaks more easily and can ignite at lower oxygen concentrations, so the hydrogen distribution and storage infrastructure will need to be airtight. Work on hydrogen tanks for vehicles has already produced coatings that can withstand pressures up to 700 kilograms per square centimeter.

Probably the best way to secure SuperCables is to run them through tunnels deep underground. Burial could significantly reduce public and political opposition to the construction of new lines.

The costs of tunneling are high, but they have been falling as underground construction and microtunneling have made great strides, as demonstrated by New York City's Water Tunnel Number 3 and the giant storm sewers in Chicago. Automated boring machines are now digging a 10.4-kilometer-long, 14.4-meter-diameter hydroelectric tunnel beside the Niagara River, at a cost of $600 million. Recent studies at Fermilab estimated the price of an 800-kilometer-long, three-meter-wide, 150-meter-deep tunnel at less than $1,000 a meter.

SuperCables would carry many times the power of existing transmission lines, which helps the economic case for burial. But the potential for further technology innovation and the limits imposed by the economics of underground construction need more exploration.

To jump-start the SuperGrid, and to clarify the costs, participants in the 2004 SuperGrid workshop proposed constructing a one-kilometer-long SuperCable to carry several hundred megawatts. This first segment would simply test the superconducting components, using liquid nitrogen to cool them. The project could be sponsored by the DOE, built at a suitable national laboratory site, and overseen by a consortium of electric utilities and regional transmission operators. Success on that prototype should lead to a 30- to 80-kilometer demonstration project that relieves real bottlenecks on today's grid by supplementing chronically congested interties between adjacent regional grids.

Beyond that, price may largely determine whether any country will muster the political and social will to construct a SuperGrid. The investment will undoubtedly be enormous: perhaps $1 trillion in today's dollars and in any case beyond the timescale attractive to private investment. It is difficult to estimate the cost of a multidecade, multigenerational SuperGrid effort. But one can judge the ultimate benefits: a carbonless, ecologically gentle domestic energy infrastructure yielding economic and physical security.

—JULY 2006

High Hopes for Hydrogen

Using hydrogen to fuel cars may eventually slash oil consumption and carbon emissions, but it will take some time.

JOAN OGDEN

Overview

- Hydrogen fuel-cell cars could become commercially feasible if automakers succeed in developing safe, inexpensive, durable models that can travel long distances before refueling.
- Energy companies could produce large amounts of hydrogen at prices competitive with gasoline, but building the infrastructure of distribution will be costly.

Developing cleaner power sources for transportation is perhaps the trickiest piece of the energy puzzle. The difficulty stems from two discouraging facts. First, the number of vehicles worldwide, now 750 million, is expected to triple by 2050, thanks largely to the expanding buying power of customers in China, India and other rapidly developing countries. And second, 97 percent of transportation fuel currently comes from crude oil.

In the near term, improving fuel economy is the best way to slow the rise in oil use and greenhouse gas emissions from cars and trucks. But even if automakers triple the efficiency of their fleets and governments support mass transit and smart-growth strategies that lessen the public's reliance on cars, the explosive growth in the number of vehicles around the world will severely limit any reductions in oil consumption and carbon dioxide emissions. To make deeper cuts, the transportation sector needs to switch to low-carbon, nonpetroleum fuels. Liquid fuels derived from woody plants or synthesized from tar sands or coal may play important roles. Over the long term, however, the most feasible ways to power vehicles with high efficiency and zero emissions are through connections to the electric grid or the use of hydrogen as a transportation fuel.

Unfortunately, the commercialization of electric vehicles has been stymied by a daunting obstacle: even large arrays of batteries cannot store enough charge to keep cars running for distances comparable to gasoline engines. For this reason, most auto companies have abandoned the technology. In contrast, fuel-cell vehicles—which combine hydrogen fuel and oxygen from the air to generate the power to run electric motors—face fewer technical hurdles and have the enthusiastic support of auto manufacturers, energy companies and policymakers. Fuel-cell vehicles are several times as efficient as today's conventional gasoline cars, and their only tailpipe emission is water vapor.

What is more, hydrogen fuel can be made without adding any greenhouse gases to the atmosphere. For example, the power needed to produce hydrogen from electrolysis—using electricity to split water into hydrogen and oxygen—can come from renewable energy sources such as solar cells, wind turbines, hydroelectric plants and geothermal facilities. Alternatively, hydrogen can be extracted from fossil fuels such as natural gas and coal, and the carbon by-products can be captured and sequestered underground.

Before a hydrogen-fueled future can become a reality, however, many complex challenges must be overcome. Carmakers must learn to manufacture

new types of vehicles, and consumers must find them attractive enough to buy. Energy companies must adopt cleaner techniques for producing hydrogen and build a new fuel infrastructure that will eventually replace the existing systems for refining and distributing gasoline. Hydrogen will not fix all our problems tomorrow; in fact, it could be decades before it starts to reduce greenhouse gas emissions and oil use on a global scale. It is important to recognize that a hydrogen transition will be a marathon, not a sprint.

The Fuel-Cell Future

Over the past decade, 17 countries have announced national programs to develop hydrogen energy, committing billions of dollars in public funds. In North America more than 30 U.S. states and several Canadian provinces are developing similar plans. Most major car companies are demonstrating prototype hydrogen vehicles and investing hundreds of millions of dollars into R&D efforts. Honda, Toyota and General Motors have announced plans to commercialize fuel-cell vehicles sometime between 2010 and 2020. Automakers and energy companies such as Shell, Chevron and BP are working with governments to introduce the first fleets of hydrogen vehicles, along with small refueling networks in California, the northeastern U.S., Europe and China.

The surge of interest in hydrogen stems not only from its long-term environmental benefits but also from its potential to stimulate innovation. Auto manufacturers have embraced fuel-cell cars because they promise to become a superior consumer product. The technology offers quiet operation, rapid acceleration and low maintenance costs. Replacing internal-combustion engines with fuel cells and electric motors eliminates the need for many mechanical and hydraulic subsystems; this change gives automakers more flexibility in designing these cars and the ability to manufacture them more efficiently. What is more, fuel-cell vehicles could provide their owners with a mobile source of electricity that might be used for recreational or business purposes. During periods of peak power usage, when electricity is most expensive, fuel-cell cars could also act as distributed generators, providing relatively cheap supplemental power for offices or homes while parked nearby.

Automakers, however, must address several technical and cost issues to make fuel-cell cars more appealing to consumers. A key component of the automotive fuel cell is the proton-exchange membrane (PEM), which separates the hydrogen fuel from the oxygen. On one side of the membrane, a catalyst splits the hydrogen atoms into protons and electrons; then the protons cross the membrane and combine with oxygen atoms on the other side. Manufacturers have reduced the weight and volume of PEM fuel cells so that they easily fit inside a compact car. But the membranes degrade with use—current automotive PEM fuel cells last only about 2,000 hours, less than half the 5,000-hour lifetime needed for commercial vehicles. Companies are developing more durable membranes, however, and in late 2005 researchers at 3M, the corporation best known for Scotch tape and Post-it notes, reported new designs that might take fuel cells to 4,000 hours and beyond within the next five years.

Another big challenge is reducing the expense of the fuel cells. Today's fuel-cell cars are hand-made specialty items that cost about $1 million apiece. Part of the reason for the expense is the small scale of the test fleets; if fuel-cell cars were mass-produced, the cost of their propulsion systems would most likely drop to a more manageable $6,000 to $10,000. That price is equivalent to $125 per kilowatt of engine power, which is about four times as high as the $30-per-kilowatt cost of a comparable internal-combustion engine. Fuel cells may require new materials and manufacturing methods to reach parity with gasoline engines. Car companies may also be able to lower costs by creatively redesigning the vehicles to fit the unique characteristics of the fuel cell. GM officials have stated that fuel-cell cars might ultimately become less expensive than gasoline vehicles because they would have fewer moving parts and a more flexible architecture.

Automobile engineers must also figure out how to store enough hydrogen in a fuel-cell car to ensure a reasonable driving range—say, 300 miles. Storing hydrogen in its gaseous state requires large, high-pressure cylinders. Although liquid hydrogen takes up less space, it must be supercooled to temperatures below −253 degrees Celsius (−423 degrees Fahrenheit). Automakers are exploring the use of metal hydride systems that adsorb hydrogen under pressure, but these devices tend to be heavy (about 300 kilograms). Finding a better storage method is a major thrust of hydrogen R&D worldwide. In the absence of a breakthrough technology, most fuel-cell vehicles today opt for the simplicity of storing the hydrogen as a compressed gas. With clever packaging and increased pressure, these cars are approaching viable travel ranges without compromising trunk space or vehicle weight. In 2005 GM, Honda and Toyota demonstrated compact fuel-cell cars with a 300-mile range using hydrogen gas compressed at 70 megapascals. (Atmospheric pressure at sea level is about 0.1 megapascal.)

Finally, safety is a necessary precondition for introducing any new fuel. Although hydrogen is flammable, it has a higher ignition temperature than gasoline and disperses in the air much more quickly, reducing the risk of fire. On the downside, a much wider range of concentrations of hydrogen is flammable, and a hydrogen flame is barely visible. Oil refineries, chemical plants and other industrial facilities already handle vast quantities of hydrogen without incident, and with proper engineering it can be made safe for consumer applications as well. The U.S. Department of Energy and other groups are currently developing safety codes and standards for hydrogen fuel.

Once hydrogen cars are introduced, how soon could they capture a large share of the market and start to significantly reduce carbon emissions and oil use? Because cars last about 15 years, it would take at least that long to switch over the entire fleet. Typically after a new automotive technology undergoes precommercial research, development and demonstration, it is introduced to the market in a single car model and only later appears in a variety of vehicles. (For example, hybrid gas-electric engines were first developed for compact sedans and later incorporated into SUVs.) Costs generally fall as production volumes increase, making the innovation more attractive. It can take 25 to 60 years for a new technology to penetrate a sizable fraction of the fleet. Although fundamental research on hybrid vehicles began in the 1970s, it was not until 1993 that Toyota began development of the Prius hybrid. Initial sales began in late 1997, but eight years later hybrid models from several manufacturers still accounted for only 1.2 percent of new vehicle sales in the U.S.

Harvesting Hydrogen

Like electricity, hydrogen must be produced from some energy source. Currently, the vast majority of hydrogen is obtained from the high-temperature processing of natural gas and petroleum. Oil refineries use hydrogen to purify petroleum-derived fuels, and chemical manufacturers employ the gas to make ammonia and other compounds. Hydrogen production now consumes 2 percent of global energy, and its share is growing rapidly. If all this hydrogen were devoted to fuel-cell cars, it would power about 150 million vehicles, or about 20 percent of the world's fleet. Although most hydrogen is produced and immediately used inside refineries or

chemical plants, some 5 to 10 percent is delivered to distant locations by truck or pipeline. In the U.S. this delivery system carries enough energy to fuel several million cars, and it could serve as a springboard to a hydrogen economy.

Making hydrogen from fossil fuels, however, generates carbon dioxide as a by-product. If hydrogen were produced from natural gas, the most common method today, and used in an efficient fuel-cell car, the total greenhouse gas emissions would work out to be about 110 grams per kilometer driven. This amount is somewhat less than the total emissions from a gasoline hybrid vehicle (150 grams per kilometer) and significantly less than those from today's conventional gasoline cars (195 grams per kilometer).

The ultimate goal, though, is to produce hydrogen with little or no greenhouse gas emissions. One option is to capture the carbon dioxide emitted when extracting hydrogen from fossil fuels and inject it deep underground or into the ocean. This process could enable large-scale, clean production of hydrogen at relatively low cost, but establishing the technical feasibility and environmental safety of carbon sequestration will be crucial. Another idea is biomass gasification—heating organic materials such as wood and crop wastes so that they release hydrogen and carbon monoxide. (This technique does not add greenhouse gases to the atmosphere, because the carbon emissions are offset by the carbon dioxide absorbed by the plants when they were growing.) A third possibility is the electrolysis of water using power generated by renewable energy sources such as wind turbines or solar cells.

Although electrolysis and biomass gasification face no major technical hurdles, the current costs for producing hydrogen using these methods are high: $6 to $10 per kilogram. (A kilogram of hydrogen has about the same energy content as a gallon of gasoline, but it will propel a car several times as far because fuel cells are more efficient than conventional gasoline engines.) According to a recent assessment by the National Research Council and the National Academy of Engineering, however, future technologies and large-scale production and distribution could lower the price of hydrogen at the pump to $2 to $4 per kilogram. In this scenario, hydrogen in a fuel-cell car would cost less per kilometer than gasoline in a conventional car today.

Nuclear energy could also provide the power for electrolysis, although producing hydrogen this way would not be significantly cheaper than using renewable sources. In addition, nuclear plants could generate hydrogen without electrolysis: the intense heat of the reactors can split water in a thermochemical reaction. This process might produce hydrogen more cheaply, but its feasibility has not yet been proved. Moreover, any option involving nuclear power has the same drawbacks that have dogged the nuclear electric power industry for decades: the problems of radioactive waste, proliferation and public acceptance.

A New Energy Infrastructure

Because the U.S. has such rich resources of wind, solar and biomass energy, making large amounts of clean, inexpensive hydrogen will not be so difficult. The bigger problem is logistics: how to deliver hydrogen cheaply to many dispersed sites. The U.S. currently has only about 100 small refueling stations for hydrogen, set up for demonstration purposes. In contrast, the country has 170,000 gasoline stations. These stations cannot be easily converted to hydrogen; the gas is stored and handled differently from liquid fuels such as gasoline, requiring alternative technologies at the pump.

The need for a new infrastructure has created a "chicken and egg" problem for the incipient hydrogen economy. Consumers will not buy hydrogen vehicles unless fuel is widely available at a reasonable price, and fuel suppliers will not build hydrogen stations unless there are enough cars to use them. And although the National Research Council's study projects that hydrogen will become competitive with gasoline once a large distribution system is in place, hydrogen might cost much more during the early years of the transition.

One strategy for jump-starting the changeover is to first focus on fleet vehicles—local delivery vans, buses and trucks—that do not require an extensive refueling network. Marine engines and locomotives could also run on hydrogen, which would eliminate significant emissions of air pollutants. Hydrogen fuel cells might power small vehicles that now use electric batteries, such as forklifts, scooters and electric bikes. And fuel cells could also be used in stationary power production: for example, they could generate electricity for police stations, military bases and other customers that do not want to rely solely on the power grid. These niche markets could help bring down the cost of fuel cells and encourage energy companies to build the first commercial hydrogen stations.

To make a substantial dent in global oil use and greenhouse gas emissions, however, hydrogen fuel will have to succeed in passenger vehicle markets. Researchers at the University of California, Davis, have concluded that 5 to 10 percent of urban service stations (plus a few stations connecting cities) must offer hydrogen to give fuel-cell car owners roughly the same convenience enjoyed by gasoline customers. GM has estimated that providing national coverage for the first million hydrogen vehicles in the U.S. would require some 12,000 hydrogen stations in cities and along interstates, each costing about $1 million. Building a full-scale hydrogen system serving 100 million cars in the U.S. might cost several hundred billion dollars, spent over decades. This estimate counts not only the expense of building refueling stations but also the new production and delivery systems that will be needed if hydrogen becomes a popular fuel.

Those numbers may sound daunting, but the World Energy Council projects that the infrastructure costs of maintaining and expanding the North American gasoline economy over the next 30 years will total $1.3 trillion, more than half of which will be spent in oil-producing countries in the developing world. Most of these costs would go toward oil exploration and production. About $300 billion would be for oil refineries, pipelines and tankers—facilities that could eventually be replaced by a hydrogen production and delivery system. Building a hydrogen economy is costly, but so is business as usual.

Furthermore, there are several ways to deliver hydrogen to vehicles. Hydrogen can be produced regionally in large plants, then stored as a liquid or compressed gas, and distributed to refueling stations by truck or gas pipeline. It is also possible to make

SEE *Figure 8 in color section.*

hydrogen locally at stations—or even in homes—from natural gas or electricity. In the early stages of a hydrogen economy, when the number of fuel-cell vehicles is relatively small, truck delivery or on-site production at refueling stations might be the most economical options. But once a large hydrogen demand is established—say, 25 percent of all the cars in a large city—a regional centralized plant with pipeline delivery offers the lowest cost. Centralized hydrogen production also opens the way for carbon sequestration, which makes sense only at large scales.

In many respects, hydrogen is more like electricity than gasoline. Because hydrogen is more costly to store and transport than gasoline, energy companies will most likely produce the fuel all over the country, with each generation plant serving a

regional market. What is more, the supply pathways will vary with location. A hydrogen economy in Ohio—which has plentiful coal and many suitable sites for carbon dioxide sequestration—might look entirely different from one in the Pacific

SEE *Figure 9 in color section.*

Northwest (which has low-cost hydropower) or one in the Midwest (which can rely on wind power and biofuels). A small town or rural area might rely on truck delivery or on-site production, whereas a large, densely populated city might use a pipeline network to transport hydrogen.

Developing a hydrogen economy will certainly entail some financial risks. If an energy company builds giant production or distribution facilities and the fuel-cell market grows more slowly than expected, the company may not be able to recoup its investments. This dilemma is sometimes called the "stranded asset" problem. The energy industry can minimize its risk, though, by adding hydrogen supply in small increments that closely follow demand. For example, companies could build power plants that generate both electricity and a small stream of hydrogen for the early fuel-cell cars. To distribute the hydrogen, the companies could initially use truck delivery and defer big investments such as pipelines until a large, established demand is in place.

The First Steps

The road to a hydrogen transportation system actually consists of several parallel tracks. Raising fuel economy is the essential first step. Developing lightweight cars, more efficient engines and hybrid electric drivetrains can greatly reduce carbon emissions and oil use over the next few decades. Hydrogen and fuel cells will build on this technical progression, taking advantage of the efficiency improvements and the increasing electrification of the vehicles.

The development of the hydrogen fuel infrastructure will be a decades-long process moving in concert with the growing market for fuel-cell vehicles. Through projects such as the California Hydrogen Highways Network and HyWays in Europe, energy companies are already providing hydrogen to test fleets and demonstrating refueling technologies. To enable fuel-cell vehicles to enter mass markets in 10 to 15 years, hydrogen fuel must be widely available at a competitive price by then. Concentrating hydrogen projects in key regions such as southern California or the Northeast corridor might help hasten the growth of the fuel-cell market and reduce the cost of infrastructure investments.

In the near term, the bulk of the hydrogen fuel will most likely be extracted from natural gas. Fueling vehicles this way will cut greenhouse gas emissions only modestly compared with driving gasoline hybrids; to realize hydrogen's full benefits, energy companies must either make the gas from zero-carbon energy sources or sequester the carbon by-products. Once hydrogen becomes a major fuel—say, in 2025 or beyond—governments should phase in requirements for zero or near-zero emissions in its production. And in the meantime, policymakers should encourage the ongoing efforts to develop clean-power technologies such as wind, solar, biomass gasification and carbon sequestration. The shift to a hydrogen economy can be seen as part of a broader move toward lower-carbon energy.

Although the transition may take several decades, hydrogen fuel-cell vehicles could eventually help protect the global climate and reduce America's reliance on foreign oil. The vast potential of this new industry underscores the importance of researching, developing and demonstrating hydrogen technologies now, so they will be ready when we need them.

—SEPTEMBER 2006

TRANSPORTATION OPTIONS

Liquid Fuels from Natural Gas

*Natural gas is cleaner and more plentiful than oil.
New ways to convert it to liquid form may soon make it
just as cheap and convenient to use in vehicles.*

SAFAA A. FOUDA

Recently countless California motorists have begun contributing to a remarkable transition. Few of these drivers realize that they are doing something special when they tank up their diesel vehicles at the filling station. But in fact, they are helping to wean America from crude oil by buying a fuel made in part from natural gas.

Diesel fuel produced in this unconventional way is on sale in California because the gas from which it is derived is largely free of sulfur, nitrogen and heavy metals—substances that leave the tailpipe as noxious pollutants. Blends of ordinary diesel fuel and diesel synthesized from natural gas (currently produced commercially by Shell in Indonesia) meet the toughest emissions standards imposed by the California Air Resources Board.

But natural gas is not only the cleanest of fossil fuels, it is also one of the most plentiful. Industry analysts estimate that the world holds enough readily recoverable natural gas to produce 500 billion barrels of synthetic crude—more than twice the amount of oil ever found in the U.S. Perhaps double that quantity of gas can be found in coal seams and in formations that release gas only slowly. Thus, liquid fuels derived from natural gas could keep overall production on the rise for about a decade after conventional supplies of crude oil begin to dwindle.

Although global stocks of natural gas are enormous, many of the deposits lie far from the people in need of energy. Yet sending gas over long distances often turns out to be prohibitively expensive. Natural gas costs four times as much as crude oil to transport through pipelines because it has a much lower energy density. The so-called stranded gas can be cooled and compressed into a liquid for shipping by tanker. Unfortunately, the conversion facilities required are large and complex, and because liquefied natural gas is hard to handle, the demand for it is rather limited.

But what if there were a cheap way to convert natural gas to a form that remains liquid at room temperature and pressure? Doing so would allow the energy to be piped to markets inexpensively. If the liquid happened to be a fuel that worked in existing vehicles, it could substitute for oil-based gasoline and diesel. And oil producers would stand to profit in many instances by selling liquid fuels or other valuable chemicals made using the gas coming from their wells.

Right now the gas released from oil wells in many parts of the world holds so little value that it is either burned on site or reinjected into the ground. In Alaska alone, oil companies pump about 200 million cubic meters (roughly seven billion cubic feet) of natural gas back into the ground daily—in large part to avoid burdening the atmosphere with additional carbon dioxide, a worrisome greenhouse gas.

But recent technical advances have prompted several oil companies to consider building plants to convert this natural gas into liquid form, which could then be delivered economically through the Alaska pipeline. On the Arabian Peninsula, the nation of Qatar is negotiating with three petrochemical companies to build gas conversion plants that would exploit a huge offshore field—a single reservoir that contains about a tenth of the world's proved gas reserves. And Norway's largest oil company, Statoil, is looking at building relatively small modules mounted on floating platforms to transform gas in remote North Sea fields into liquids. Although these efforts will use somewhat different technologies, they all must address the same fundamental problem in chemistry: making larger hydrocarbon molecules from smaller ones.

The Classic Formula

The main component of natural gas is methane, a simple molecule that has four hydrogen atoms neatly arrayed around one carbon atom. This symmetry makes methane particularly stable. Converting it to a liquid fuel requires first breaking its chemical bonds. High temperatures and pressures help to tear these bonds apart. So do cleverly designed catalysts, substances that can foster a chemical reaction without themselves being consumed.

The conventional "indirect" approach for converting natural gas to liquid form relies on brute force. First, the chemical bonds in methane are broken using steam, heat and a nickel-based catalyst to produce a mixture of carbon monoxide and hydrogen known as syngas (or, more formally, synthesis gas). This process is called steam reforming.

The second step in the production of liquid fuels (or other valuable petrochemicals) from syngas uses a method invented in 1923 by Franz Fischer and Hans Tropsch. During World War II, Germany harnessed this technique to produce liquid fuels using syngas made from coal and atmospheric oxygen, thus establishing a reliable internal source for gasoline and diesel.

This Fischer-Tropsch technology has allowed Sasol in South Africa to produce liquid fuels commercially for decades using syngas derived from coal. The company uses the same basic technique today: syngas blown over a catalyst made of cobalt, nickel or iron transforms into various liquid hydrocarbons. Conveniently, the Fischer-Tropsch

Color Section

Figure 1 (see page 11)

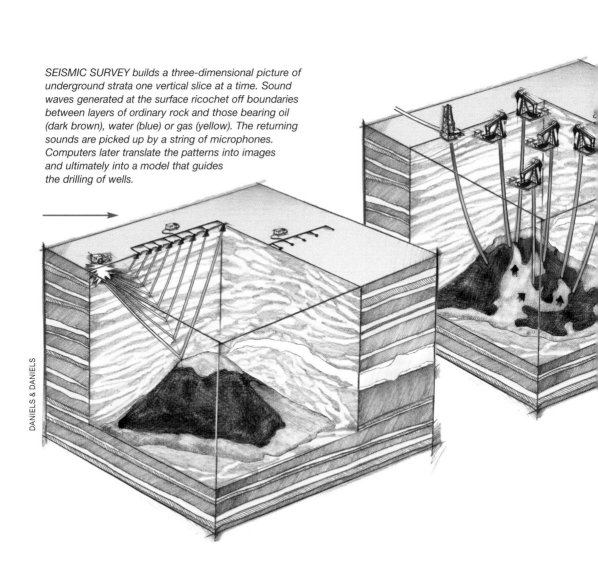

SEISMIC SURVEY builds a three-dimensional picture of underground strata one vertical slice at a time. Sound waves generated at the surface ricochet off boundaries between layers of ordinary rock and those bearing oil (dark brown), water (blue) or gas (yellow). The returning sounds are picked up by a string of microphones. Computers later translate the patterns into images and ultimately into a model that guides the drilling of wells.

DANIELS & DANIELS

INJECTION OF LIQUID CARBON DIOXIDE can rejuvenate dying oil fields. Pumped at high pressure from tanks into wells that have ceased producing oil, the carbon dioxide flows through the reservoir and, if all goes well, pushes the remaining oil down toward active wells. Steam and natural gas are sometimes also used for this purpose. Alternatively, water can be injected below a pocket of bypassed crude in order to shepherd the oil into a well. In the future, "smart" wells currently under development will be able to retrieve oil simultaneously from some branches of the well while using other branches to pump water out of the oil stream and back into the formation from which it came.

PRODUCTION WELLS often draw water from below and gas from above into pore spaces once full of oil. This complex flow strands pockets of crude far from wells; traditional drilling techniques thus miss up to two thirds of the oil in a reservoir. But repeated seismic surveys can now be assembled into a 4-D model that not only tracks where oil, gas and water in the field are located but also predicts where they will go next. Advanced seismic monitoring works well on about half the world's oil fields, but it fails on oil buried in very hard rock or beneath beds of salt (thick white layer).

Figure 2 (see page 12)

HORIZONTAL DRILLING was impractical when oil rigs had to rotate the entire drill string—up to 5,800 meters (roughly 19,000 feet) of it—in order to turn the rock-cutting bit at the bottom. Wells that swing 90 degrees over a space of just 100 meters are now common thanks to the development of motors that can run deep underground. The motor's driveshaft connects to the bit through a transmission in a bent section of pipe. The amount of bend determines how tight a curve the drill will carve; drillers can twist the string to control the direction of the turn.

DANIELS & DANIELS (background illustration)

SENSORS near the bit can detect oil, water and gas. One device measures the porosity of the surrounding rock by emitting neutrons, which scatter off hydrogen atoms. Another takes a density reading by shooting out gamma rays that interact with adjacent electrons. Oil and water also affect electrical resistance, measured from a current passed through the bit, the rock and nearby electrodes.

LAURIE GRACE; SOURCE: SCHLUMBERGER

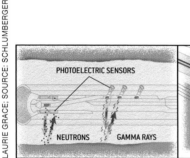

PHOTOELECTRIC SENSORS

NEUTRONS   GAMMA RAYS

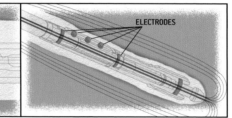

ELECTRODES

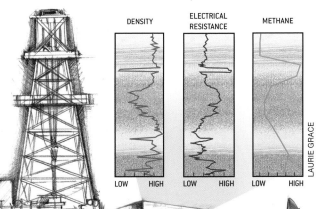

DENSITY   ELECTRICAL RESISTANCE   METHANE

LOW HIGH   LOW HIGH   LOW HIGH

LAURIE GRACE

GEOLOGIC MEASUREMENTS collected by sensors near the bottom of the drill pipe can be analyzed at the well-head or transmitted via satellite to engineers anywhere in the world. Several characteristics of the rocks surrounding the drill bit can reveal the presence of oil or gas (left). Petroleum tends to accumulate in relatively light, porous rocks, for example, so some geosteering systems calculate the bulk density of nearby strata. Others measure the electrical resistance of the earth around the drill; layers soaked with briny water have a much lower resistance than those rich in oil. Gas chromatographs at the surface analyze the returning flow of lubricating mud for natural gas captured during its journey.

"SMART" WELLS of the near future will use computers and water monitors near the bottom of the well to detect dilution of the oil stream by water. Hydrocyclonic separators will then shunt the water into a separate branch of the well that empties beneath the oil reservoir.

a

b

ADVANCED DRILLS use mud pumped through the inside of the string to rotate the bit, to commu-nicate sensor measurements and to carry rock fragments out of the well. On its way down, the mud first enters a rotating valve (a), which converts data radioed to the tool from various sen-sors into surges in the mud stream. (At the surface, the pulses are translated back into a digital signal of up to 10 bits per second.) The mud next flows into a motor. A spiral driveshaft fits inside the helical motor casing in a way that creates chambers (b). As the cavities fill with mud, the shaft turns in order to relieve the hydraulic pressure. The mud finally exits through the rotating bit and returns to the surface, with fresh cuttings cleared from near the bit.

FORKED WELLS can extract oil from several oil-bearing layers at once. Computer-controlled chokes inserted in the well pipe maintain the optimum flow of oil to the surface.

ILLUSTRATION NOT TO SCALE

HIBERNIA

RAM-POWEL

DANIELS & DANIELS

THREE NEW WAYS
to tap oil fields that lie
deep underwater have
recently been deployed. Hibernia
(left), which began producing oil last
November from a field in 80 meters of water
off the coast of Newfoundland, Canada, took
seven years and more than $4 billion to construct. Its
base, built from 450,000 tons of reinforced concrete, is
designed to withstand the impact of a million-ton iceberg.
Hibernia is expected to recover 615 million barrels of oil over 18
years, using water and gas injection. Storage tanks will hold up to 1.3
million barrels of oil inside the base until it can be transferred to shuttle
tankers. Most deepwater platforms send the oil back to shore through
subsea pipelines.

DEEPEST OIL WELL in active production (right) currently lies more than 1,709 meters
beneath the waves of the South Atlantic Ocean, in the Marlim field off the coast of
Campos, Brazil. The southern part of this field alone is thought to contain 10.6 billion
barrels of oil. Such resources were out of reach until recently. Now remotely operated
submarines are being used to construct production facilities on the sea bottom itself.
The oil can then be piped to a shallower platform if one is nearby. Or, as in the case of
the record-holding South Marlim 3B well, a ship can store the oil until shuttle tankers
arrive. The challenge is to hold the ship steady above the well. Moorings can provide
stability at depths up to about 1,500 meters. Beyond that limit, ships may have to use
automatic thrusters linked to the Global Positioning System and beacons on the
seafloor to actively maintain their positions. These techniques may allow the industry
to exploit oil fields under more than 3,000 meters of water in the near future.

Figure 3 (see page 12)

RAM-POWELL platform (center), built by Shell Oil, Amoco and Exxon, began production in the Gulf of Mexico last September. The 46-story platform is anchored to 270-ton piles driven into the seafloor 980 meters below. Twelve tendons, each 71 centimeters in diameter, provide a strong enough mooring to withstand 22-meter waves and hurricane winds up to 225 kilometers per hour. The $1-billion facility can sink wells up to six kilometers into the seabed in order to tap the 125 million barrels of recoverable oil estimated to lie in the field. A 30-centimeter pipeline will transport the oil to platforms in shallower water 40 kilometers away. Ram-Powell is the third such tension leg platform completed by Shell in three years. Next year, Shell's plans call for an even larger platform, named Ursa, to start pumping 2.5 times as much oil as Ram-Powell from below 1,226 meters of water.

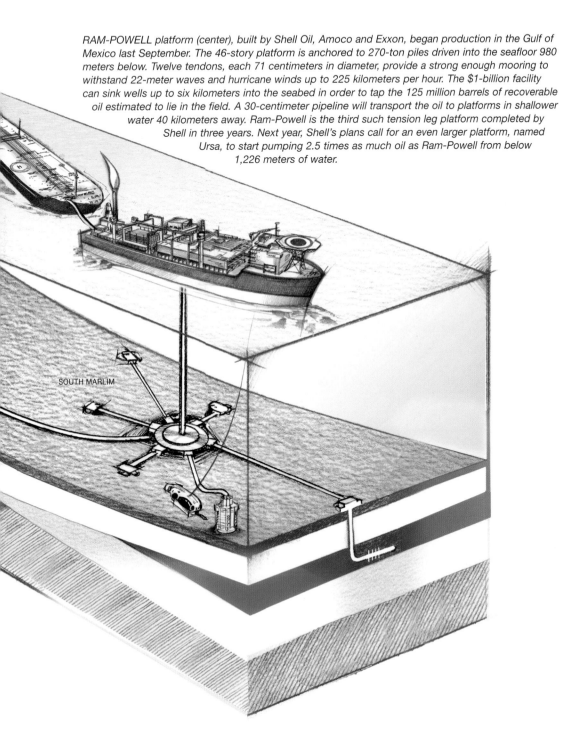

SOUTH MARLIM

Figure 4 (see page 35)

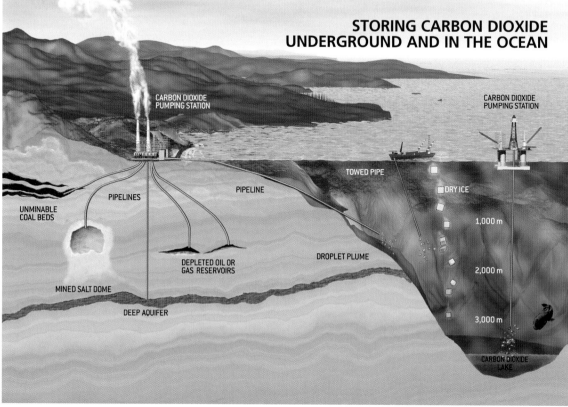

STORING CARBON DIOXIDE UNDERGROUND AND IN THE OCEAN

CARBON DIOXIDE PUMPING STATION

CARBON DIOXIDE PUMPING STATION

TOWED PIPE

DRY ICE

1,000 m

PIPELINES

PIPELINE

UNMINABLE COAL BEDS

2,000 m

DEPLETED OIL OR GAS RESERVOIRS

DROPLET PLUME

MINED SALT DOME

3,000 m

DEEP AQUIFER

CARBON DIOXIDE LAKE

DAVID FIERSTEIN

STORAGE UNDERGROUND	ADVANTAGES	DISADVANTAGES	STORAGE IN OCEAN	ADVANTAGES	DISADVANTAGES
Coal Beds	Potentially low costs	Immature technology	Droplet Plume	Minimal environmental effects	Some leakage
Mined Salt Domes	Custom designs	High costs	Towed Pipe	Minimal environmental effects	Some leakage
Deep Saline Aquifers	Large capacity	Unknown storage integrity	Dry Ice	Simple technology	High costs
Depleted Oil or Gas Reservoirs	Proven storage integrity	Limited capacity	Carbon Dioxide Lake	Carbon will remain in ocean for thousands of years	Immature technology

STORAGE SITES for carbon dioxide in the ground and deep sea should help keep the greenhouse gas out of the atmosphere where it now contributes to climate change. The various options must be scrutinized for cost, safety and potential environmental effects.

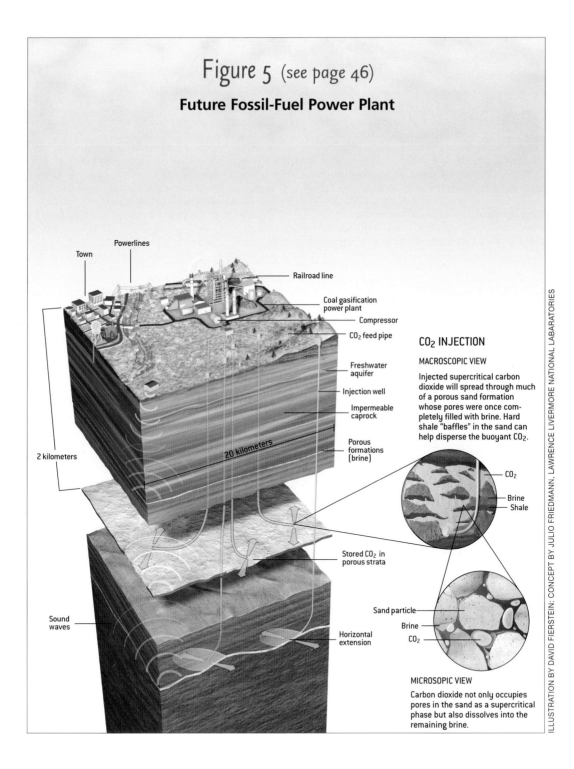

Figure 5 (see page 46)

Future Fossil-Fuel Power Plant

Town

Powerlines

Railroad line

Coal gasification
power plant

Compressor

CO_2 feed pipe

Freshwater
aquifer

Injection well

Impermeable
caprock

Porous
formations
(brine)

20 kilometers

2 kilometers

Stored CO_2 in
porous strata

Sound
waves

Horizontal
extension

CO_2 INJECTION

MACROSCOPIC VIEW

Injected supercritical carbon
dioxide will spread through much
of a porous sand formation
whose pores were once completely filled with brine. Hard
shale "baffles" in the sand can
help disperse the buoyant CO_2.

CO_2

Brine
Shale

Sand particle

Brine

CO_2

MICROSOPIC VIEW

Carbon dioxide not only occupies
pores in the sand as a supercritical
phase but also dissolves into the
remaining brine.

Figure 6 (see page 68)

The Path to CO$_2$ Mitigation

Our calculations indicate that a prompt commitment to carbon capture and storage (CCS) would make it possible to meet global energy demands while limiting the atmospheric carbon dioxide concentration to 450 parts per million by volume (ppmv). This goal could be attained if, by midcentury, sequestration is applied for all coal use and about a quarter of natural gas use, while energy efficiency increases rapidly and carbon-free energy sources expand sevenfold. Under these conditions, overall fossil-fuel consumption could expand modestly from today: by midcentury, coal use could be somewhat higher than at present,

oil use would be down by a fifth and natural gas use would expand by half.

To realize this pathway, growth rates for fossil-fuel use would have to be reduced now, and CCS must begin for coal early in the next decade and for natural gas early in the next quarter of a century. The top graph below depicts the energy provided by the various sources if this mitigation path were followed. The bottom graph shows total quantities of carbon extracted from the earth (emissions plus storage).

—D.G.H., D.A.L. and R.H.W.

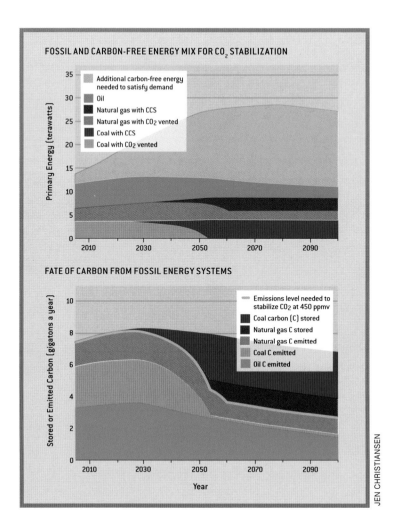

Figure 7 (see page 101)

New Type of Nuclear Reactor

A safer, more sustainable nuclear power cycle could be based on the advanced liquid-metal reactor (ALMR) design developed in the 1980s by researchers at Argonne National Laboratory. Like all atomic power plants, an ALMR-based system would use nuclear chain reactions in the core to produce the heat needed to generate electricity.

Current commercial nuclear plants feature thermal reactors, which rely on relatively slow moving neutrons to propagate chain reactions in uranium and plutonium fuel. An ALMR-based system, in contrast, would use fast-moving (energetic) neutrons. This process permits all the uranium and heavier atoms to be consumed, thereby allowing vastly more of the fuel's energy to be captured. In the near term, the new reactor would burn fuel made by recycling spent fuel from thermal reactors.

In most thermal-reactor designs, water floods the core to slow (moderate) neutrons and keep it cool. The ALMR, however, employs a pool of circu-

lating liquid sodium as the coolant (1). Engineers chose sodium because it does not slow down fast neutrons substantially and because it conducts heat very well, which improves the efficiency of heat delivery to the electric generation facility.

A fast reactor would work like this: Nuclear fire burning in the core would heat the radioactive liquid sodium running through it. Some of the heated sodium would be pumped into an intermediate heat exchanger (2), where it would transfer its thermal energy to nonradioactive liquid sodium flowing through the adjacent but separate pipes (3) of a secondary sodium loop. The nonradioactive sodium (4) would in turn bring heat to a final heat exchanger/steam generator (not shown), where steam would be created in adjacent water-filled pipes. The hot, high-pressure steam would then be used to turn steam turbines that would drive electricity-producing generators (not shown).

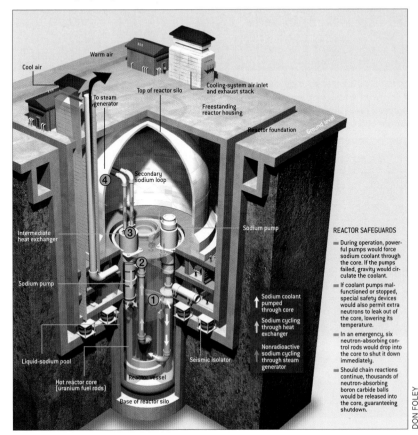

Figure 8 (see page 145)

Options for a Hydrogen Infrastructure

Energy companies could manufacture and distribute hydrogen fuel in many ways. In the near term, the most likely option is extracting hydrogen from natural gas, either in centralized reformers that supply fueling stations by delivery truck or in smaller on-site reformers located at the stations. The fueling stations could also use electricity from the power grid to make hydrogen by electrolyzing water. All these options, however, would produce greenhouse gas emissions (assuming that fossil fuels are used to make the electricity).

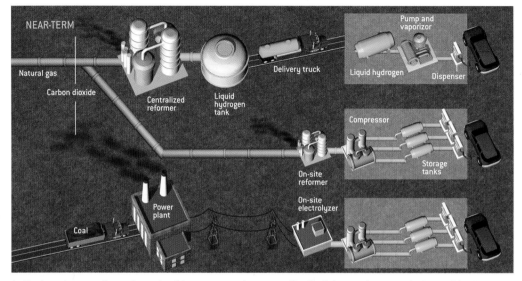

In the long term, policymakers should encourage cleaner methods. Advanced power plants could extract hydrogen from coal and bury the carbon dioxide deep underground. Wind turbines and other renewable energy sources could provide the power for electrolysis. And high-temperature steam from nuclear reactors could generate hydrogen through the thermochemical splitting of water.

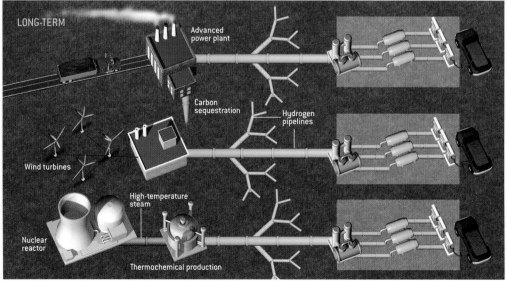

DON FOLEY

Figure 9 (see page 145)

The Many Uses of Hydrogen

Because transporting hydrogen over long distances would be costly, each generation plant would serve the surrounding region. The first users would most likely include fleet vehicles such as trucks and small vehicles that now use electric batteries (for example, forklifts at a warehouse). Hydrogen fuel cells could also power marine engines and provide supplemental electricity for office buildings. The owners of fuel-cell cars could stop at hydrogen stations or even generate their own hydrogen at home using the power from solar arrays.

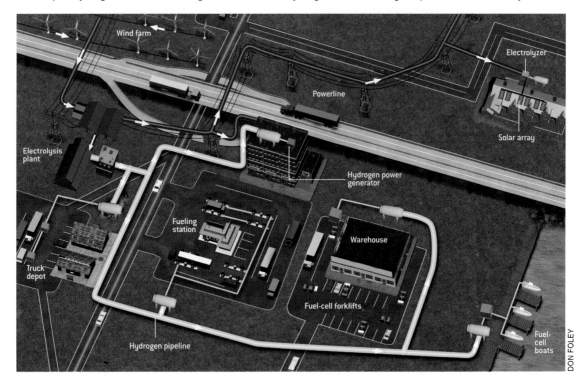

Figure 10 (see page 155)

FUEL-CELL POWER PLANT

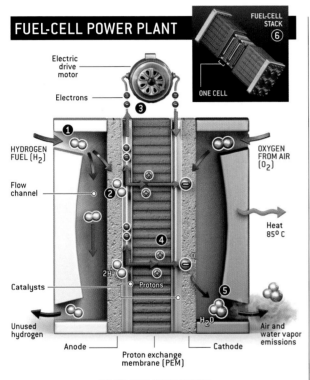

FUEL-CELL STACK (6)

ONE CELL

Electric drive motor

Electrons

(3)

(1)

HYDROGEN FUEL (H_2)

Flow channel

(2)

OXYGEN FROM AIR (O_2)

Heat 85° C

(4)

2H

Catalysts

Protons

(5)

H_2O

Unused hydrogen

Air and water vapor emissions

Anode

Proton exchange membrane (PEM)

Cathode

UP TO 55% EFFICIENCY

ELECTROCHEMISTRY VS. COMBUSTION: A proton exchange membrane (PEM) fuel cell comprises two thin, porous electrodes, an anode and a cathode, separated by a polymer membrane electrolyte that passes only protons. Catalysts coat one side of each electrode. After hydrogen enters (1), the anode catalyst splits it into electrons and protons (2). The electrons travel off to power a drive motor (3), while the protons migrate through the membrane (4) to the cathode. Its catalyst combines the protons with returning electrons and oxygen from the air to form water (5). Cells can be stacked to provide higher voltages (6).

INTERNAL-COMBUSTION ENGINE

FUEL AND AIR MIXTURE

Spark plug

Exhaust valve

Intake valve

Exhaust emissions
• Carbon dioxide
• Nitrogen oxides
• Hydrocarbons
• Carbon monoxide
• Sulfur dioxide

Piston

Heat 125° C

Cylinder

Connecting rod

Crankshaft

JOE ZEFF

UP TO 30% EFFICIENCY

The piston, which travels up and down when the crankshaft rotates, starts at the top of the cylinder. The intake valve opens and the piston drops, allowing the fuel/air mixture to enter the cylinder. The piston moves back up, compressing the gasoline and air. The spark plug fires, igniting the fuel droplets. The compressed charge explodes, driving the piston down. The exhaust valve opens, allowing the combustion products to exit the cylinder.

Figure 11 (see page 184)

LED Performance

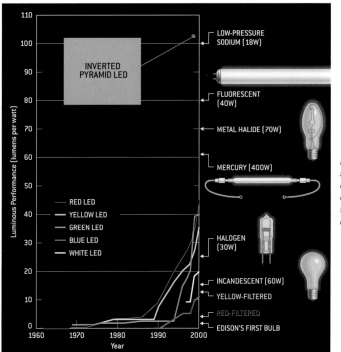

Light-emitting diodes have steadily improved and now outperform many other kinds of lights; the best is a proto-type red-orange inverted pyra-mid LED.

Semiconductor chip is the key to an LED's glow. An applied voltage drives "holes" (positive charges) from the p-type layer and electrons from the n-type layer into the active layer. When they meet, they give off photons. The color of the photons depends on the chemical makeup of the layers, although some manufacturers house LEDs in colored lenses as a means of identification (photograph).

LED: The Inside View

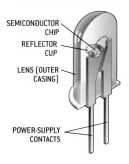

SEMICONDUCTOR CHIP
REFLECTOR CUP
LENS (OUTER CASING)
POWER-SUPPLY CONTACTS

The Heart of the LED

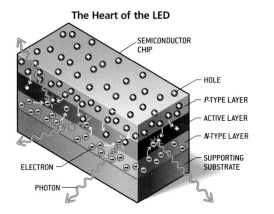

SEMICONDUCTOR CHIP
HOLE
P-TYPE LAYER
ACTIVE LAYER
N-TYPE LAYER
SUPPORTING SUBSTRATE
ELECTRON
PHOTON

Figure 12 (see page 196)

Wind Power

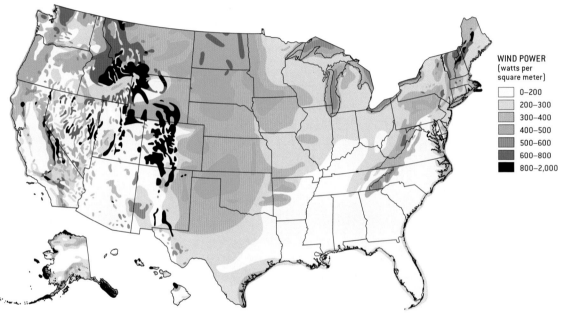

WIND POWER
(watts per
square meter)

	0–200
	200–300
	300–400
	400–500
	500–600
	600–800
	800–2,000

America has enormous wind energy resources, enough to generate as much as 11 trillion kilowatt-hours of electricity each year. Some of the best locations for wind turbines are the Great Plains states, the Great Lakes and the mountain ridges of the Rockies and the Appalachians.

reaction gives off heat, and often this heat is used to drive the oxygen compressors needed to make syngas.

Just which liquids emerge from the reaction depends on temperature. For example, running a reaction vessel at 330 to 350 degrees Celsius (626 to 662 degrees Fahrenheit) will primarily produce gasoline and olefins (building blocks often used to make plastics). A cooler (180 to 250 degrees C) operation will make predominantly diesel and waxes. In any case, a mixture results, so a third and final step is required to refine the products of the reaction into usable fuels.

Refining synthetic crudes derived from gas is in many respects easier than working with natural crude oil. Synthetic crude contains virtually no sulfur and has smaller amounts of cancer-causing compounds than are found in conventional oil. So the final products are premium-quality fuels that emit fewer harmful substances.

A Partial Solution

This brute-force method of converting gas to liquids is reliable, but it is expensive because it uses so much energy. Conventional steam reforming compresses methane and water vapor to about 30 times normal atmospheric pressure and heats these reactants to about 900 degrees C. And one must add more heat still, to coax the energy-hungry reaction continuously along. This extra heat comes from injecting a small amount of oxygen into the mixture, which combusts some of the methane (and, as an added benefit, makes more syngas). Chemists call this latter maneuver partial oxidation.

In general, syngas is generated using various combinations of steam reforming and partial oxidation. In most cases, the process requires large quantities of oxygen—and oxygen is costly. Existing methods of separating oxygen from air rely on refrigeration to cool and liquefy it, an energy-intensive and expensive manipulation.

Hence, lowering the cost of oxygen is the key to making syngas cheaply.

Fortunately, recent developments promise to revolutionize the way oxygen is produced over the next few years. One strategy is simply to work with air instead of pure oxygen. Syntroleum Corporation in Tulsa has developed a way to make liquid fuels using blown air and methane for the reforming step, followed by Fischer-Tropsch synthesis. At sites where natural gas is sufficiently cheap (for example, places where it is now being flared), the process should prove profitable even at current crude oil prices. Together with Texaco and the English company Brown & Root, Syntroleum plans to build a commercial plant that will use this technique within two years.

Several other private companies, universities and government research laboratories are pursuing a wholly different approach to the oxygen problem: they are developing ceramic membranes through which only oxygen can pass. These membranes can then serve as filters to purify oxygen from air. Though still difficult and expensive to construct, laboratory versions work quite well. They should be commercially available within a decade.

Such materials could reduce the cost of making syngas by about 25 percent and lower the cost of producing liquid fuels by 15 percent. These savings would accrue because the production of syngas could be done at temperatures about 200 degrees lower than those currently used and because there would be no need to liquefy air. With cheap and plentiful oxygen, partial oxidation alone could supply syngas. This first step would then release energy rather than consume it.

My Canadian colleagues and I, along with researchers at the University of Florida, are now attempting to create a different kind of ceramic membrane that would offer yet another advantage. The membranes we are trying to develop would remove hydrogen from the gas mixture, driving the partial oxidation of methane forward and providing a stream of pure hydrogen that could be used later in refining the final products or as an energy source itself.

We also expect to see significant improvements soon in the catalysts used to make syngas. In particular, researchers at the University of Oxford are studying metal carbides, and my colleagues at the Canadian Center for Mineral and Energy Technology are investigating large-pore zeolites. Both materials show great promise in reducing the soot generated during operation, a problem that not only plugs the reactor but also reduces the activity of the catalysts over time.

Cheaper than Oil?

Although the prospects for such brute-force methods of converting natural gas to liquid fuel improve every day, more ingenious techniques on the horizon would accomplish that transformation in a single step. This approach could potentially cut the cost of conversion in half, which would make liquid fuels produced from natural gas actually less expensive than similar products refined from crude oil.

Early efforts to achieve such "direct" conversion by using different catalysts and adding greater amounts of oxygen had produced mostly disappointment. The hydrocarbons that were formed proved more reactive than the methane supplied. In essence, they burned up faster than they were produced. Unless the product is somehow removed from the reaction zone, yields are too low to be practical.

Fortunately, researchers have recently found ways to circumvent this problem. The trick is to run the reaction at comparatively mild temperatures using exotic catalysts or to stabilize the product chemically—or to do both. For example, chemists at Pennsylvania State University have converted

methane to methanol directly using a so-called homogeneous catalyst, a liquid that is thoroughly mixed with the reactants and held at temperatures lower than 100 degrees C. And Catalytica, a company in Mountain View, California, has achieved yields for direct conversion that are as high as 70 percent using a similar scheme. Its liquid catalyst creates a relatively stable chemical intermediate, methyl ester, that is protected from oxidation. The final product (a methanol derivative) is easily generated with one subsequent step.

Methanol (also known as wood alcohol) is valuable because it can be readily converted to gasoline or to an octane-boosting additive. And in the near future methanol (either used directly or transformed first into hydrogen gas) could also serve to power fuel-cell vehicles on a wide scale. Thus, methanol can be regarded as a convenient currency for storing and transporting energy.

Moreover, the reactions used to synthesize methanol can be readily adjusted to churn out diesel alternatives such as dimethyl ether, which produces far fewer troublesome pollutants when it burns. So far dimethyl ether, like propane, has found little use as a transportation fuel because it is a gas at room temperature and pressure. But recently Air Products, a supplier of industrial gases in Allentown, Pennsylvania, announced the production of a dimethyl ether derivative that is liquid at ambient conditions. So this substitute for conventional diesel fuel would reduce emissions without major changes to vehicles and fueling stations.

Now You're Cooking with Gas

Scientists and engineers are pursuing many other possible ways to improve the conversion of natural gas into liquids. For instance, process developers are constantly improving the vessels for the Fischer-Tropsch reaction to provide better control of heat and mixing.

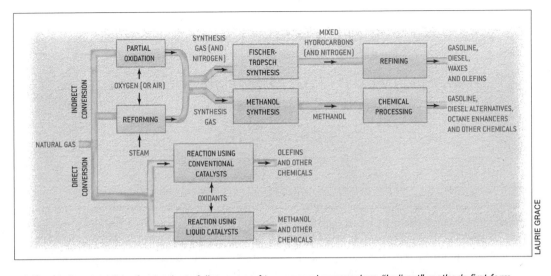

Chemical conversion of natural gas follows one of two general approaches. "Indirect" methods first form synthesis gas, or syngas (a mixture of carbon monoxide and hydrogen), by adding steam and oxygen or oxygen alone. (One company, Syntroleum, uses air instead, removing the unwanted nitrogen during later processing.) The second step synthesizes larger molecules from syngas, and the third step refines or chemically tailors the various products. "Direct" conversion of natural gas in one step requires an oxidant and may involve special liquid catalysts.

The most ambitious efforts now under way attempt to mimic the chemical reactions used by specialized bacteria that consume methane in the presence of oxygen to produce methanol. Low-temperature biological reactions of this kind are quite promising because they can produce specific chemicals using relatively little energy.

Whether or not this bold line of research ultimately succeeds, it is clear that even today natural gas can be converted into liquid fuels at prices that are only about 10 percent higher per barrel than crude oil. Modest improvements in technology, along with the improved economics that come from making specialty chemicals as well from gas, will broaden the exploitation of this abundant commodity in coming years. Such developments will also provide remarkably clean fuels—ones that can be easily blended with dirtier products refined from heavier crude oils to meet increasingly strict environmental standards. So the benefits to society will surely multiply as people come to realize that natural gas can do much more than just run the kitchen stove.

—MARCH 1998

Pumping Coal

Coming soon to the U.S.: cleaner diesel from dirty coal.

GUNJAN SINHA

The U.S. is plump with coal. The country has one-quarter of the world's reserves, and coal accounts for about 50 percent of the nation's electricity. To cut the reliance on oil imports, why not also use it to power cars and trucks or to heat homes, too?

That may happen soon. This year Waste Management and Processors, Inc. (WMPI) will break ground for the first U.S. coal-to-diesel production facility, in Gilberton, Pennsylvania. The plant will process 1.4 million tons of waste coal a year to generate approximately 5,000 barrels a day of diesel fuel. Other states, such as Illinois, Virginia, Kentucky, Wyoming and West Virginia, are also considering coal-to-liquid facilities.

Interest in the technology is certainly welcome news to WMPI president John Rich, who has been trying to finance such a facility for more than a decade. "Coal to liquids hadn't taken off, because the price of crude was at $30 to $40 a barrel," Rich says. Oil at about $60 makes coal more attractive.

To create the fuel, coal is first mixed with oxygen and steam at high temperature and pressure to produce carbon monoxide and hydrogen. The second step, referred to as Fischer-Tropsch synthesis, uses a catalyst to transform the gas into a liquid synthetic crude, which is further refined. Along the way, mercury, sulfur, ammonia and other compounds are extracted for sale on the commodities market.

The type of technology required to gasify the coal depends on the starting material. Pennsylvania alone has an estimated 260 million tons of waste coal—coal discarded because of its low energy content. "For every two tons of coal mined, up to half ends up in the reject pile," Rich says. Existing nearby facilities are not equipped to burn it. WMPI will rely on approaches innovated by South African energy giant Sasol; those methods are optimized to work with energy-poor coal, which includes lignite and bitumen.

The resultant fuel is cleaner than conventional, sulfur-free diesel. In comparison tests, Daimler-Chrysler showed that the coal-derived fuel spews 10 percent of the carbon monoxide and hydrocarbons and 70 percent of the particulates. The firm had plans to unveil a demonstration vehicle with a tweaked V-6 engine in April that cuts nitrogen oxides and other emissions even further, says Stefan Keppeler, senior manager of fuels research at the company.

Though relatively clean at the tailpipe, the fuel is dirty at its source. A similar coal-based power plant discharges about four million tons of carbon dioxide a year. In some facilities, the greenhouse gas can be repurposed—it can be pumped into oil fields or, in the case of WMPI's plant, sold to the beverage industry. Unless scientists develop methods to sequester CO_2 and find other uses for the gas, the technology might languish, warns Rudi Heydenrich, business unit manager at Sasol. The gasification step is also expensive, accounting for two-thirds of the cost of a facility. "You need a structure where there is government support to ensure sustainable economics in the long run," Heydenrich remarks.

Under the Bush administration's Clean Coal Power Initiative, a $100 million federal loan guarantee jump-started the new WMPI facility. The state of Pennsylvania also chipped in with tax credits and a plan to buy up to half the plant's output to power its vehicles. Investors may contribute the additional $500 million necessary to build the plant. The initial cost of the fuel is expected to be about $54 a barrel.

Coal is not the only source of synthetic diesel; the fuel can be derived from natural gas and more cheaply, too. In fact, Qatar and Nigeria are building gas-to-liquid plants, and Sasol estimates that by 2014, gas-to-liquid fuel may account for at least 5 percent of the global market. But the U.S. does not have nearly as much natural gas as coal. And considering the vast coal reserves in China, which is also considering the technology, coal-derived diesel seems likely to play a bigger role in helping to liberate some countries from dependence on oil imports.

—MAY 2006

Vehicle of Change

Hydrogen fuel-cell cars could be the catalyst for a cleaner tomorrow.

LAWRENCE D. BURNS, J. BYRON McCORMICK AND CHRISTOPHER E. BORRONI-BIRD

When Karl Benz rolled his Patent Motorcar out of the barn in 1886, he literally set the wheels of change in motion. The advent of the automobile led to dramatic alterations in people's way of life as well as the global economy—transformations that no one expected at the time. The ever-increasing availability of economical personal transportation remade the world into a more accessible place while spawning a complex industrial infrastructure that shaped modern society.

Now another revolution could be sparked by automotive technology: one fueled by hydrogen rather than petroleum. Fuel cells—which cleave hydrogen atoms into protons and electrons that drive electric motors while emitting nothing worse than water vapor—could make the automobile much more environmentally friendly. Not only could cars become cleaner, they could also become safer, more comfortable, more personalized—and even perhaps less expensive. Further, these fuel-cell vehicles could be instrumental in motivating a shift toward a "greener" energy economy based on hydrogen. As that occurs, energy use and production could change significantly. Thus, hydrogen fuel-cell cars and trucks could help ensure a future in which personal mobility—the freedom to travel

99 percent, carbon monoxide by 96 percent and nitrogen oxides by 95 percent—the continued production of carbon dioxide causes concern because of its potential to change the planet's climate.

Even with the application of new technologies, the efficiency of the petroleum-fueled ICE is expected to plateau around 30 percent—and whatever happens, it will still discharge carbon dioxide. In comparison, the hydrogen fuel-cell vehicle is nearly twice as efficient, so it will require just half the fuel energy. Of even more significance, fuel cells emit only water and heat as byproducts. Finally, hydrogen gas can be extracted from various fuels and energy sources, such as natural gas, ethanol, water (via electrolysis using electricity) and, eventually, renewable energy systems. Realizing this potential, an impressive roster of automotive companies is making a sustained effort to develop fuel-cell vehicles, including Daimler-Chrysler, Ford, General Motors, Honda, PSA Peugeot-Citroën, Renault-Nissan and Toyota.

It's an Automotive World

It is important to find a better solution to the problems posed by personal transportation because the environmental impact of vehicles is apt to wax as use booms. In 1960 fewer than 4 percent of the world's population possessed vehicles. Twenty years later 9 percent were owners, and currently the share has reached 12 percent. Based on present growth rates, as many as 15 percent of the people living on the planet could have a vehicle by 2020. And because the world's population may climb from six billion today to nearly 7.5 billion two decades hence, the total number of vehicles could increase from about 700 million to more than 1.1 billion. This projected expansion will be spurred by the burgeoning of the middle class in the developing world, which translates into rising per capita income. Higher income correlates almost directly with automobile ownership.

independently—could be sustained indefinitely, without compromising the environment or depleting the earth's natural resources.

A confluence of factors makes the big change seem increasingly likely. For one, the petroleum-fueled internal-combustion engine (ICE), as highly refined, reliable and economical as it is, is finally reaching its limits. Despite steady improvements, today's ICE vehicles are only 20 to 25 percent efficient in converting the energy content of fuels into drive-wheel power. And although the U.S. auto industry has cut exhaust emissions substantially since the unregulated 1960s—hydrocarbons dropped by

Three-quarters of all automobiles are now concentrated in the U.S., Europe and Japan. We expect, however, that more than 60 percent of the increase in new vehicle sales during the next 10 years will occur in eight emerging markets: China, Brazil, India, Korea, Russia, Mexico, Poland and Thailand. The challenge will be to create compelling, affordable and profitable vehicles that are safe, effective and environmentally sustainable.

Rethinking Automotive Propulsion

To understand why this technology could be so revolutionary, consider the operation of a fuel-cell vehicle, which at base is a vehicle with an electric traction drive. Instead of an electrochemical battery, though, the motor gets power from a fuel-cell unit. Electricity is produced when electrons are stripped from hydrogen fuel traveling through a membrane in the cell. The resulting current runs the electric motor, which turns the wheels. The hydrogen protons then combine with oxygen and electrons to form water. When using pure hydrogen, a fuel-cell car is a zero-emission vehicle.

SEE *Figure 10 in color section.*

Although it takes energy to extract hydrogen from substances, by either reforming hydrocarbon molecules with catalysts or splitting water with electricity, the fuel cell's high efficiency more than compensates for the energy required to accomplish these processes, as we will show later. Of course, this energy has to come from somewhere. Some generation sources, such as natural gas-, oil- and coal-burning power facilities, produce carbon dioxide and other greenhouse gases. Others, including nuclear plants, do not. An optimal goal would be to produce electricity from renewable sources such as biomass, hydroelectric, solar, wind or geothermal energy.

By adopting hydrogen as an automotive fuel, the transportation industry could begin the transition from near-total reliance on petroleum to a mix of fuel sources. Today 98 percent of the energy used to power automobiles is derived from petroleum. As a result, roughly two-thirds of the oil imported into the U.S. is devoted to transportation. By supplementing fossil fuels, the U.S. can reduce dependence on foreign oil and foster development of local, more environmentally friendly energy sources. This effort will also introduce competition into energy pricing—which could lower and stabilize fuel and energy costs in the long term.

Revamping Vehicle Design

Another key to producing a truly revolutionary vehicle is the integration of the fuel cell with drive-by-wire technology, replacing previous, predominantly mechanical systems for steering, braking, throttling and other functions with electronically controlled units. This frees up space because electronic systems tend to be less bulky than mechanical ones. By-wire system performance can be programmed using software. In addition, with no conventional drivetrain to limit structural and styling choices, automakers will be free to create dramatically different designs to satisfy customer needs.

Replacing conventional ICEs with fuel cells enables the use of a flat chassis, which gives designers great freedom to create unique body styles. Drive-by-wire technology similarly liberates the interior design because the driving controls can be radically altered and can be operated from different seating positions. Recognizing this design opportunity, General Motors came up with a concept called AUTOnomy, which the company introduced early this year. A drivable prototype, Hy-wire (for hydrogen by-wire), debuted at the Paris Motor Show in late September.

The AUTOnomy concept and the Hy-wire prototype were created, literally, from the wheels up. The foundation for both is a thin, skateboard-like chassis containing the fuel cell, electric drive motor, hydrogen storage tanks, electronic controls and heat exchangers, as well as braking and steering systems [see sidebar, "AUTOnomy's 'Skateboard' Chassis," below]. There is no internal-combustion engine, transmission, drivetrain, axles or mechanical linkages.

In a fully developed AUTOnomy-type vehicle, drive-by-wire technology would require only one simple electrical connection and a set of mechanical links to unite chassis and body. The body could plug into the chassis much like a laptop connects to a docking station. The single-electrical-port concept creates a quick and easy way to link all the body systems—controls, power and heating—to the skateboard. This simple separation of body and chassis can help keep the vehicle body lightweight and uncomplicated. It also makes the body easily replaceable. In principle, simply by having the dealer or car owner "pop on" an interchangeable body module, the vehicle could be a luxury car today, a family sedan next week or a minivan next year.

Much like a computer, vehicle systems would be upgradable through software. As a result, service personnel could download programs as desired to improve vehicle performance or to tailor particular ride and handling characteristics to suit a particular vehicle brand, body style or customer preference.

With drive-by-wire electronic controls, the driver needs no steering wheel, gear shifter or foot pedals. GM's Hy-wire prototype is equipped with a steering guide control called X-Drive that easily moves from side to side across the width of the car to accommodate left- and right-hand driving positions. The X-Drive operates something like a motorcycle's handgrips: the driver accelerates by twisting the handgrips and brakes by squeezing them. Steering involves a turning action similar to today's steering wheel. The driver also has the option to brake and accelerate with either the right or left hand, with braking taking priority in the case of mixed signals. Motorists start the vehicle by pushing a single power button and then select one of three settings: neutral, drive or reverse.

X-Drive also eliminates the conventional instrument panel and steering column, which frees up the vehicle interior and allows novel placement of seats and storage areas. For example, because there is no engine compartment, the driver and all passengers have more visibility and much greater legroom than in a conventional vehicle of the same length.

By lowering the vehicle's center of gravity and eliminating the rigid engine block in front of the

AUTOnomy's "Skateboard" Chassis

Shoehorning functional automotive systems into the flat, skateboard-like chassis is the key to General Motors AUTOnomy concept for a future hydrogen fuel-cell vehicle. That and the use of compact electronic drive-by-wire technology for steering, braking and throttling permits designers much greater freedom in configuring the upper bodies. It means no more bulky engine compartment, awkward center cabin hump or conventional steering wheel to work around. The novel approach also allows bodies to be interchangeable. Owners could have new, personalized bodies "plugged in" to their used chassis at the dealership, or do it themselves—turning, say, a family sedan into a minivan or a luxury car.

passengers, an AUTOnomy-like skateboard chassis can improve ride, handling and stability characteristics beyond what is possible with conventional vehicle architecture.

Reorganizing the Automobile Business

The simplified design of an advanced fuel-cell vehicle, as suggested by the AUTOnomy concept and the Hy-wire prototype, could have a profound effect on vehicle manufacturing, perhaps setting the stage for a reinvention of the automobile business. Today's auto industry is capital intensive, with modest profit margins. Even as car companies are aggressively managing the costs of vehicle development and manufacturing, excess production capacity in the global industry is driving down vehicle prices. At the same time, the regulatory standards–driven content of cars and trucks continues to grow, pushing up costs. Taken together, lower prices and higher costs are threatening profit margins.

A concept such as that of AUTOnomy, however, could significantly change the current business model. It could conceivably lower vehicle development costs because, with modules able to be produced independently, design changes to the body and chassis modules could be made more easily and cheaply. As with today's truck platform derivatives, it will be possible to design the chassis only once to accommodate various body styles. These derivatives could easily have different front ends, interior layouts and chassis tuning. With perhaps only three chassis needed—compact, mid-size and large—production volumes could be much larger than those now, bringing greater economies of scale.

Having far fewer components and part types will further reduce costs. The fuel-cell stack, for example, is created from a series of identical individual cells, each comprising a flat cathode sheet

and similar anode component separated by a polymer-electrolyte membrane. Depending on the power requirements of a particular vehicle (or other device, such as a stationary electricity generator), the number of cells in the stack can be scaled up or down.

Although automotive fuel-cell technology is far from economical at present (thousands of dollars per kilowatt for a hand-built prototype), costs have begun to decline dramatically. For instance, the 10-fold increase in the power density of fuel-cell stacks achieved over the past five years has been accompanied by a 10-fold decrease in their cost. And whereas fuel cells currently require precious metals for catalysts and expensive polymer membranes, scientists are making progress in finding ways to minimize the use of catalysts and make membrane materials cheaper.

The AUTOnomy concept also makes it possible to decouple body and chassis manufacture. A global manufacturer could build and ship the chassis (an ideal scenario, given its thin profile), and local firms could build the bodies and assemble the complete vehicles. The chassis could be very economical because it would be mass-produced.

In high-end markets, this kind of arrangement might mean that new chassis might debut every three or four years—when software upgrades could no longer match performance desires—but that customers could purchase a new body module annually or lease one even more frequently. In addition, if chassis hardware is developed appropriately, then new hardware and software upgrades could become practical. Alternatively, consumers who wish to keep their vehicle body but want a higher-performance chassis could buy one. In less affluent markets, the chassis would comprise durable hardware and could be financed for much longer periods, perhaps decades.

Steps Toward a Hydrogen Society

Within a Few Years

A Small numbers of prototype vehicles are tested by leasing them to residents living near a hydrogen fueling station.

B Transit and business fleets that return to the garage each day, such as buses, mail trucks and delivery vans, start to be supplied by centrally located hydrogen stations.

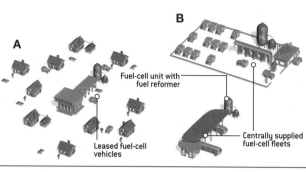

A

B

Fuel-cell unit with fuel reformer

Leased fuel-cell vehicles

Centrally supplied fuel-cell fleets

Within a Decade

A Car plants manufacture fuel-cell-powered "skateboards" and a few different "snap-on" body types.

B Hydrogen fueling stations with on-site natural gas reformers (chemical cracking units) are installed to provide hydrogen to early production vehicles.

C Stationary power generators, which reform natural gas into hydrogen and feed into the fuel cell, are installed in enterprises that require high-reliability "premium" power for data communications, continuous manufacturing, or emergency medicine. For example, ambulances and emergency vehicles refuel at the hospital fuel-cell unit.

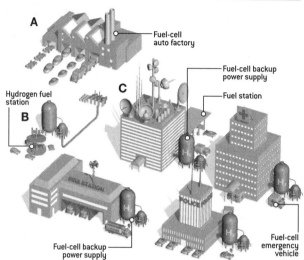

A

Fuel-cell auto factory

Fuel-cell backup power supply

Hydrogen fuel station

B

C

Fuel station

Fuel-cell backup power supply

Fuel-cell emergency vehicle

In More Than a Decade

A Stationary reformer/fuel-cell units sited at more types of businesses and, eventually, homes sell extra power to the electricity grid in what's called a distributed generation system. These installations begin to provide hydrogen locally to employees.

B More hydrogen stations that use electrolyzers come online.

C Huge assembly plants put out three sizes of fuel-cell skateboard chassis (compact, midsize, large).

D Dealers sell new bodies in various styles for drivers' used skateboards.

E Other plants in different regions build diverse bodies for their local markets (for example, in India and China, tractors and trucks).

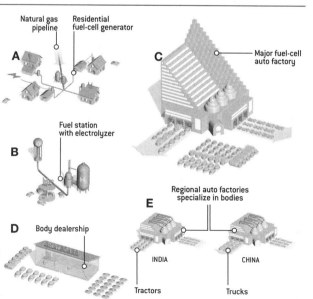

Natural gas pipeline

Residential fuel-cell generator

A

Major fuel-cell auto factory

C

Fuel station with electrolyzer

B

Body dealership

D

Regional auto factories specialize in bodies

E

INDIA

CHINA

Tractors

Trucks

Energy Sources

Current fossil-fuel, nuclear and hydroelectric generation will be increasingly augmented by cleaner and renewable technologies.

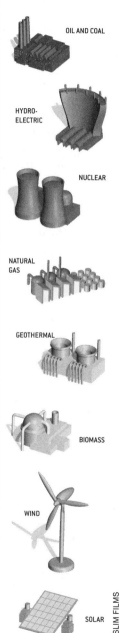

OIL AND COAL

HYDRO-ELECTRIC

NUCLEAR

NATURAL GAS

GEOTHERMAL

BIOMASS

WIND

SOLAR

SLIM FILMS

Storing Hydrogen

This is not to say that all the technical barriers to engineering practical fuel-cell vehicles have been surmounted. Many obstacles have yet to be overcome before they will achieve the convenience and performance that customers have come to expect from their ICE automobiles. One of the biggest hurdles is the development of safe and effective onboard hydrogen storage technology that would provide sufficient driving range—about 300 miles. Any acceptable storage technology must be durable enough to run for at least 150,000 miles. It must function in temperatures from –40 to 113 degrees Fahrenheit (–40 to 45 degrees Celsius). And the refueling process must be simple and take less than five minutes to complete. There are various approaches to storing hydrogen, including liquid, compressed gas and solid-state methods. All are promising, yet all present challenges.

Compressed-gas tanks are the most likely to be used early on, but high compression remains a perceived safety issue. Currently, these systems carry about 5,000 pounds per square inch (psi) of hydrogen (350 bars), but the goal is 10,000 psi (700 bars) to extend vehicle range. For safety purposes, the tank must have an impact burst strength of at least twice the pressure of the fuel. Vessels are currently made from materials that are either very expensive, such as carbon fiber, or very heavy. They are also relatively large, making it difficult to fit them in a vehicle.

Hydrogen can also be stored in liquid form, but a substantial amount of energy is needed to chill it to the extremely low temperatures required (–423 degrees F or –253 degrees C). Further, as much as 3 to 4 percent of the hydrogen will still "boil off" every day. Although most of this boil-off will be used by the vehicle, it would be a concern for cars parked for several days between trips.

A longer-term solution is to transport hydrogen using a solid-state approach. One promising alternative is metal-hydride storage. In this method, the hydrogen is held in the interstices of pressed metallic alloy powder, much like a sponge absorbs water. This technique has many encouraging aspects, including straightforward construction, a high degree of safety and promising storage capacities. But temperatures in the range of 150 to 300 degrees C are needed to extract the hydrogen from the metal hydride. To avoid an energy penalty, the hydrogen must be released at a temperature nearer to 80 degrees C. Although research is still in the early stages, solid-state storage is tantalizing.

Reworking the Infrastructure

Momentous as the changes to the automotive business might be, they could be overshadowed by the potential influence of AUTOnomy-type vehicles on the world's energy supply system. Viewed from where we are today, fuel cells and a hydrogen-fueling infrastructure are a chicken-and-egg problem. We cannot have large numbers of fuel-cell vehicles without adequate fuel availability to support them, but we will not be able to create the required infrastructure unless there are significant numbers of fuel-cell vehicles on the roadways. Given that the creation of a potentially costly hydrogen generation/distribution network in the U.S. is a prerequisite to commercializing fuel-cell cars and trucks, strong advocacy from local and national leaders in the public and private sectors is crucial. Key issues that must be addressed include subsidy funding, incentives for developing refueling stations, creation of uniform standards, and general education about the topic. The Freedom-CAR initiative announced by the U.S. Department of Energy earlier this year, a public-private partnership to promote the development of fuel-cell power and hydrogen as a primary fuel for cars and trucks, is a step in the right direction. Government support for the research and pilot demonstrations required to prove the feasibility of the infrastructure will be needed.

To be sure, industry also must do its part to enable the difficult transition to the hydrogen economy. GM is now developing a bridge strategy that should move things along. We are working on bringing to market interim hydrogen-based fuel-cell products that will earn revenues to help offset the hundreds of millions of dollars that the company is investing on fuel-cell technology, while providing real-world operating experience.

It is likely that fuel-cell generators will be marketed for use in businesses and, eventually, homes, before fuel-cell vehicles are widely available. These applications are much less complex than automobiles, which have very demanding performance requirements. GM has developed prototype stationary fuel-cell generators that run on hydrogen extracted from fossil fuels.

Within the next few years, GM plans to unveil a range of stationary fuel-cell generators that are aimed at the "premium power," or high-reliability "guaranteed power," energy market segment. This $10-billion-a-year business encompasses energy consumers that cannot afford to be without electricity, including digital-data centers, hospitals, factories using continuous industrial processes, and telecommunications companies. These generators would enable cost reduction through the ability to cut power usage during peak periods as well as provide revenue through net metering (selling power back to the grid). One of our initial products will be a 75-kilowatt unit incorporating a reformer that extracts hydrogen for the fuel-cell stack from natural gas, methane or gasoline. No breakthrough technical developments are needed to build these stationary power products. When operational, these decentralized power systems can also be used to refuel vehicles with hydrogen.

Once safe and reliable hydrogen storage methods are available, off-board fuel processing at the filling station becomes a viable avenue for generating the hydrogen needed for transportation. An advantage of fuel processing, of course, is that most of the infrastructure required to implement it already exists. The current petroleum-based fuel distribution network could be retrofitted by installing fuel reformers or electrolyzers right at the corner gas station, allowing local operators to generate hydrogen on the spot and pump it for their customers. With this approach, there would be no need to build new long-distance pipelines or dismantle the present automotive servicing infrastructure. As we begin the transition from petroleum to hydrogen, this might well be the optimal way to proceed.

An even more radical scheme would be to refuel at home or at work using the distribution network that currently provides natural gas to individual homes and businesses. Natural gas pipelines are as common in many areas as gasoline stations, making this infrastructure an ideal conduit for hydrogen. The natural gas could be reformed into hydrogen and then stored onboard the vehicle. Alternatively, electricity from the utility grid could produce the hydrogen. Electricity purchased during off-peak hours, such as when your car is housed overnight in a garage, might eventually be an affordable way to refuel in some locales.

As vehicle power-generation systems become more sophisticated, we see the role of the automobile within the global power grid changing. Vehicles could at some point become a new power-generation source, supplying electricity to homes and work sites. Most vehicles sit idle about 90 percent of the time, so imagine the exponential growth in power availability if the electrical grid could be supplemented by the generating capacity of cars and trucks in every driveway or parking garage. Consider, for example, that if only one out of every 25 vehicles in California today were a fuel-cell vehicle, their combined generating capacity would exceed that of the state's utility grid.

Obviously there are multiple options to choose from in creating a hydrogen distribution network.

Although the scenarios we have painted are plausible, one of the most important factors in determining what the infrastructure will eventually look like is cost. Energy companies around the world are studying the economics of hydrogen. In recent testimony before the U.S. House of Representatives Committee on Science's Energy Subcommittee, James P. Uihlein of BP stated that hydrogen can be generated from natural gas at a cost that is comparable with conventional fuel costs. In fact, he went on to note, at the refinery gate, hydrogen's cost-per-mile-driven is actually substantially less than conventional fuel because of the outstanding efficiency of the fuel-cell engine. Hydrogen's current high cost, Uihlein said, can be attributed to the expense of transporting and dispensing it.

Hydrogen Matters

Depending on the feedstock and the production and distribution methods used, the cost of a kilogram of hydrogen can be four to six times as high as the cost of a gallon of gasoline or diesel fuel. (A kilogram of hydrogen is the energy equivalent of a gallon of petroleum-based fuel.) Yet because an optimized fuel-cell vehicle is likely to be at least twice as efficient as an ICE vehicle, it will go twice as far on that kilogram of fuel. Therefore, hydrogen should become commercially viable if its retail price per kilogram is double that of a gallon of gasoline. As improvements in hydrogen storage, fuel processing and electrolysis technologies are achieved and as demand for hydrogen increases, the cost of hydrogen should move nearer to the required price range. In fact, recent studies indicate that with today's technology we are within a factor of 1.3 of where we would like to be in terms of price.

Even though we are in the early stages of exploring solutions, we believe that when the infrastructure is required, it could develop rapidly, despite the enormous challenges involved. That was the case a century ago, when the gasoline automobile was proving its usefulness to customers, and the infrastructure to support it grew quickly. Entrepreneurs are always ready to seize new opportunities. The world is already beginning to develop the technologies needed for hydrogen production and distribution. Nevertheless, the size and scope of this particular infrastructure are huge, and there are significant technical obstacles ahead.

As discussions about how to create the required distribution network continue, it is interesting to note that hydrogen infrastructures are currently installed in several locations, most notably along the U.S. Gulf Coast and in Europe around Rotterdam, the Netherlands. Hydrogen is produced by the oil and chemical industries (it is used for sulfur removal in the petroleum-refining process), so it flows today through hundreds of miles of pipeline in a number of countries. The existing infrastructure annually produces approximately 540 billion cubic meters of hydrogen, primarily reformed from natural gas. On an energy-equivalent basis, this equates to roughly 140 million tons of petroleum a year, which is almost 10 percent of the present transportation demand. Even though the infrastructure is dedicated to other uses, the fact that it is already in place demonstrates that a great deal of expertise on generating and transporting hydrogen is available.

Like any advance that has the potential to change the dominant technology completely, the implementation of fuel cells and the transition to a hydrogen-based energy infrastructure will take time. Although a precise timetable is hard to predict, given our current technological momentum and business realities, we aim to have compelling and affordable fuel-cell vehicles on the road by the end of this decade. We then anticipate a significant increase in the penetration of fuel-cell vehicles between 2010 and 2020 as automakers begin to create the installed base necessary to support high-volume production. Many of these companies, including GM, have invested hundreds of millions of dollars in fuel-cell research and development, and the sooner they can anticipate a return on these investments, the better.

Because it takes about 20 years to turn over the entire vehicle fleet, it will take at least that long to reap the full extent of the environmental and energy benefits that hydrogen fuel-cell vehicles can provide. But the AUTOnomy concept brings that future nearer—and makes it clearer. Instead of the historical evolution of the automobile, we now see the development of revolutionary technologies that fundamentally reinvent the automobile and its role in our world.

—OCTOBER 2002

Fueling Our Transportation Future

What are the options for decreasing demand for oil and lowering greenhouse gas emissions in cars and light trucks?

JOHN B. HEYWOOD

Overview

- The massive use of petroleum-based fuels for transportation releases immense amounts of carbon dioxide into the atmosphere—25 percent of the total worldwide.

- Options for constraining and eventually reducing these emissions include improving vehicle technology, reducing vehicle size, developing different fuels, and changing the way vehicles are used.

- To succeed, we will most likely have to follow through on all of these choices.

If we were honest, most of us in the world's richer countries would concede that we like our transportation systems. They allow us to travel when we want to, usually door-to-door, alone or with family and friends, and with baggage. The mostly unseen freight distribution network delivers our goods and supports our lifestyle. So why worry about the future and especially about how the energy that drives our transportation might be affecting our environment?

The reason is the size of these systems and their seemingly inexorable growth. They use petroleum-based fuels (gasoline and diesel) on an unimaginable scale. The carbon in these fuels is oxidized to the greenhouse gas carbon dioxide during combustion, and their massive use means that the amount of carbon dioxide entering the atmosphere is likewise immense. Transportation accounts for 25 percent of worldwide greenhouse gas emissions. As the countries in the developing world rapidly motorize, the increasing global demand for fuel will pose one of the biggest challenges to controlling the concentration of greenhouse gases in the atmosphere. The U.S. light-duty vehicle fleet (automobiles, pickup trucks, SUVs, vans and small trucks) currently consumes 150 billion gallons (550 billion liters) of gasoline a year, or 1.3 gallons of gasoline per person a day. If other nations burned gasoline at the same rate, world consumption would rise by a factor of almost 10.

As we look ahead, what possibilities do we have for making transportation much more sustainable, at an acceptable cost?

Our Options

Several options could make a substantial difference. We could improve or change vehicle technology; we could change how we use our vehicles; we could reduce the size of our vehicles; we could use different fuels. We will most likely have to do all of these to drastically reduce energy consumption and greenhouse gas emissions.

In examining these alternatives, we have to keep in mind several aspects of the existing transportation system. First, it is well suited to its primary context, the developed world. Over decades, it has had time to evolve so that it balances economic costs with users' needs and wants. Second, this vast optimized system relies completely on one convenient source of energy: petroleum. And it has evolved technologies—internal-combustion engines on land and jet engines (gas turbines) for air—that well match vehicle operation with this energy-dense liquid fuel. Finally, these vehicles last a long time. Thus, rapid change is doubly difficult. Constraining and then reducing the local and global impacts of transportation energy will take decades.

We also need to keep in mind that efficiency ratings can be misleading; what counts is the fuel consumed in actual driving. Today's gasoline spark-ignition engine is about 20 percent efficient in urban driving and 35 percent efficient at its best operating point. But many short trips with a cold engine and transmission, amplified by cold weather and aggressive driving, significantly worsen fuel consumption, as do substantial time spent with the engine idling and losses in the transmission. These real-world driving phenomena reduce the engine's average efficiency so that only about 10 percent of the chemical energy stored in the fuel tank actually drives the wheels. Amory Lovins, a strong advocate for much lighter, more efficient vehicles, has stated it this way: with a 10 percent efficient vehicle and with the driver, a passenger and luggage—a payload of some 300 pounds, about 10 percent of the vehicle weight— "only 1 percent of the fuel's energy in the vehicle tank actually moves the payload."

We must include in our accounting what it takes to produce and distribute the fuel; to drive the vehicle through its lifetime of 150,000 miles (240,000 kilometers); and to manufacture, maintain and dispose of the vehicle. These three phases of vehicle operation are often called well-to-tank (this phase accounts for about 15 percent of the total lifetime energy use and greenhouse gas emissions), tank-to-wheels (75 percent), and cradle-to-grave (10 percent). Surprisingly, the energy required to produce the fuel and the vehicle is not negligible. This total-life-cycle accounting becomes especially important as we consider fuels that do not come from petroleum and new types of vehicle technologies. It is what gets used and emitted in this *total sense* that matters.

Improving existing light-duty vehicle technology can do a lot. By investing more money in increasing the efficiency of the engine and transmission, decreasing weight, improving tires and reducing drag, we can bring down fuel consumption by about one-third over the next 20 or so years—an annual 1 to 2 percent improvement, on average. (This reduction would cost between $500 and $1,000 per vehicle; at likely future fuel prices, this amount would not increase the lifetime cost of ownership.) These types of improvements have occurred steadily over the past 25 years, but we have bought larger, heavier, faster cars and light trucks and thus have effectively traded the benefits we could have realized for these other attributes. Though most obvious in the U.S., this shift to larger, more powerful vehicles has occurred elsewhere as well. We need to find ways to motivate buyers to use the potential for reducing fuel consumption and greenhouse gas emissions to actually save fuel and contain emissions.

In the near term, if vehicle weight and size can be reduced and if both buyers and manufacturers can step off the ever-increasing horsepower/performance path, then in the developed world we may be able to slow the rate of petroleum demand, level it off in 15 to 20 years at about 20 percent above current demand, and start on a slow downward path. This projection may not seem nearly aggressive enough. It is, however, both challenging to achieve and very different from our current trajectory of steady growth in petroleum consumption at about 2 percent a year.

In the longer term, we have additional options. We could develop alternative fuels that would displace at least some petroleum. We could turn to new propulsion systems that use hydrogen or electricity. And we could go much further in designing and encouraging acceptance of smaller, lighter vehicles.

The alternative fuels option may be difficult to implement unless the alternatives are compatible with the existing distribution system. Also, our current fuels are liquids with a high-energy density: lower-density fuels will require larger fuel tanks or provide less range than today's roughly 400 miles. From this perspective, one alternative that stands out is nonconventional petroleum (oil or tar sands, heavy oil, oil shale, coal). Processing these sources to yield "oil," however, requires large amounts of other forms of energy, such as natural gas and electricity. Thus, the processes used emit substantial amounts of greenhouse gases and have other environmental impacts. Further, such processing calls for big capital investments. Nevertheless, despite the broader environmental consequences, nonconventional petroleum sources are already starting to be exploited; they are expected to provide some 10 percent of transportation fuels within the next 20 years.

Biomass-based fuels such as ethanol and biodiesel, which are often considered to emit less carbon dioxide per unit of energy, are also already being produced. In Brazil ethanol made from sugarcane constitutes some 40 percent of transport fuel. In the U.S. roughly 20 percent of the corn crop is being converted to ethanol. Much of this is blended with gasoline at the 10 percent level in so-called reformulated (cleaner-burning) gasolines. The recent U.S. national energy policy act plans to double ethanol production from the current 2 percent of transportation fuel by 2012. But the fertilizer, water and natural gas and electricity currently expended in ethanol production from corn will need to be substantially decreased. Production of ethanol from cellulosic biomass (residues and wastes from plants not generally used as a food source) promises to be more efficient and to lower greenhouse gas emissions. It is not yet a commercially viable process, although it may well become so. Biodiesel can be made from various crops (rapeseed, sunflower, soybean oils) and waste animal fats. The small amounts now being made are blended with standard diesel fuel.

It is likely that the use of biomass-based fuels will steadily grow. But given the uncertainty about the environmental impacts of large-scale conversion of biomass crops to fuel (on soil quality, water resources and overall greenhouse gas emissions), this source will contribute but is unlikely to dominate the future fuel supply anytime soon.

Use of natural gas in transportation varies around the world from less than 1 percent to 10 to 15 percent in a few countries where tax policies make it economical. In the 1990s natural gas made inroads into U.S. municipal bus fleets to achieve lower emissions; diesels with effective exhaust cleanup are now proving a cheaper option.

What about new propulsion system technology? Likely innovations would include significantly improved gasoline engines (using a turbocharger with direct fuel injection, for example), more efficient transmissions, and low-emission diesels with catalysts and particulate traps in the exhaust, and perhaps new approaches to how the fuel is combusted might be included as well. Hybrids, which combine a small gasoline engine and a battery-powered electric motor, are already on the road, and production volumes are growing. These vehicles use significantly less gasoline in urban driving, have lower benefits at highway speeds and cost a few thousand dollars extra to buy.

Researchers are exploring more radical propulsion systems and fuels, especially those that have the potential for low life-cycle carbon dioxide emissions. Several organizations are developing hydrogen-powered fuel-cell vehicles in hybrid form with a battery and an electric motor. Such systems could increase vehicle efficiency by a factor of two, but much of that benefit is offset by the energy consumed and the emissions produced in making and distributing hydrogen. If the hydrogen can be produced through low-carbon-emitting processes and if a practical distribution system could be set up, it has low-greenhouse-emissions potential. But it would take technological breakthroughs and many decades before hydrogen-based transportation could become a reality and have widespread impact.

Hydrogen is, of course, an energy carrier rather than an energy source. Electricity is an alternative energy carrier with promise of producing energy without releasing carbon dioxide, and various research teams are looking at its use in transportation. The major challenge is coming up with a battery that can store enough energy for a reasonable driving range, at an acceptable cost. One technical barrier is the long battery recharging time. Those of us used to filling a 20-gallon tank in four minutes might have to wait for several hours to charge a battery. One way around the range limitation of electric vehicles is the plug-in hybrid, which has a small engine on board to recharge the battery when needed. The energy used could thus be largely electricity and only part engine fuel. We do not yet know whether this plug-in hybrid technology will prove to be broadly attractive in the marketplace.

Beyond adopting improved propulsion systems, a switch to lighter-weight materials and different vehicle structures could reduce weight and improve fuel consumption without downsizing. Obviously, though, combining lighter materials and smaller vehicle size would produce an even greater effect. Maybe the way we use vehicles in the future will differ radically from our "general purpose vehicle" expectations of today. In the future, a car specifically designed for urban driving may make sense. Volkswagen, for example, has a small two-person concept car prototype that weighs 640 pounds (290 kilograms) and consumes one liter of gasoline per 100 kilometers (some 240 miles per gallon—existing average U.S. light-duty vehicles use 10 liters per 100 kilometers, or just under 25 miles per gallon). Some argue that downsizing reduces safety, but these issues can be minimized.

Promoting Change

Better technology will undoubtedly improve fuel efficiency. In the developed world, markets may even adopt enough of these improvements to offset the expected increases in the number of vehicles. And gasoline prices will almost certainly rise over the next decade and beyond, prompting changes in the way consumers purchase and use their vehicles. But market forces alone are unlikely to curb our ever-growing appetite for petroleum.

A coordinated package of fiscal and regulatory policies will need to come into play for fuel-reduction benefits to be realized from these future improvements. Effective policies would include a "feebate" scheme, in which customers pay an extra fee to buy big fuel-consumers but get a rebate if they buy small, fuel-efficient models. The feebate combines well with stricter Corporate Average Fuel Economy (CAFE) standards—in other words, with regulations that require automobile makers to produce products that consume less fuel. Adding higher fuel taxes to the package would further induce people to buy fuel-efficient models. And tax incentives could spur more rapid changes in the production facilities for new technologies. All these measures may be needed to keep us moving forward.

—SEPTEMBER 2006

AN ENERGY-EFFICIENT ECONOMY

A Waste of Energy

THE EDITORS

America needs a new energy policy to reduce its reliance on foreign oil, but the $26 billion measure that stalled in Congress [November 2003] clearly wasn't it. The bill was bloated with $17 billion in tax breaks intended to spur production of oil, natural gas, coal and nuclear power. Although the act would have also funded efforts to reduce greenhouse gas emissions—such as the Clean Coal Power Initiative—its strategy was wasteful and wrongheaded. The energy bill would have spent billions of taxpayer dollars on the development of unproven technologies that may never be adopted by the private sector.

Rather than resurrecting the failed 2003 bill this year, Congress should start afresh with a law focused on energy conservation. The energy saved through efficiency measures since the 1970s has been far greater than that produced by any new oil field or coal mine. As those measures came into effect between 1979 and 1986, the U.S. gross domestic product rose 20 percent while total

energy use dropped 5 percent. Last year's energy bill would have set new efficiency standards for several products (traffic signals, for instance) and provided tax incentives for energy-efficient buildings and appliances, but the government can do much more.

Many economists argue that the best conservation strategy would be to establish an across-the-board energy tax. Under this approach, Congress would not dictate any efficiency standards; rather businesses and consumers would voluntarily avoid energy-guzzling appliances, heating systems and vehicles to minimize their tax bills. European countries, for example, have successfully boosted the average fuel economy of their cars by imposing high taxes on gasoline. But raising energy taxes would place a disproportionate burden on poor Americans if the new excises were not accompanied by some relief for low-income people. And the idea is a political nonstarter in Washington, D.C., anyway.

A more palatable approach would be to bolster energy conservation efforts that are already proving their worth. More than 20 states have public benefits funds that assess small charges on electricity use (typically about a tenth of a cent per kilowatt-hour) and direct the money toward efficiency upgrades. New York's Energy Smart Program, for instance, has cut annual energy bills in the state by more than $100 million since 1998, and current projects are expected to double the savings. Nationwide, however, ratepayer-financed programs lost ground in the 1990s because of utility deregulation. Congress can correct this problem by creating a federal fund that would match the state investments.

Another smart move would be to raise the Corporate Average Fuel Economy (CAFE) standards for cars and light trucks. Thanks in large part to CAFE, which was introduced in 1975, the average gas mileage of new vehicles in the U.S. reached a high of 26.2 miles per gallon in 1987. But the average has slid to 25.1 mpg since then, partly because more people are buying sport-utility vehicles, which are held to a lower standard than cars. At the very least, Congress should remove the loophole for SUVs. Automakers have the technology to improve fuel economy, and consumers will benefit in the end because their savings at the gas pump will far outweigh any markups at the car dealership.

According to the American Council for an Energy-Efficient Economy, a law that establishes a federal benefits fund and raises CAFE standards could reduce annual energy usage in the U.S. by nearly 12 percent. To put it another way, conservation would eliminate the need to build 700 new power plants. That's a lot of juice.

—FEBRUARY 2004

More Profit with Less Carbon

Focusing on energy efficiency will do more than protect Earth's climate—it will make businesses and consumers richer.

AMORY B. LOVINS

A basic misunderstanding skews the entire climate debate. Experts on both sides claim that protecting Earth's climate will force a trade-off between the environment and the economy. According to these experts, burning less fossil fuel to slow or prevent global warming will increase the cost of meeting society's needs for energy services, which include everything from speedy transportation to hot showers. Environmentalists say the cost would be modestly higher but worth it; skeptics, including top U.S. government officials, warn that the extra expense would be prohibitive. Yet both sides are wrong. If properly done, climate protection would actually *reduce* costs, not raise them. Using energy more efficiently offers an economic bonanza—not because of the benefits of stopping global warming but because saving fossil fuel is a lot cheaper than buying it.

The world abounds with proven ways to use energy more productively, and smart businesses are leaping to exploit them. Over the past decade, chemical manufacturer DuPont has boosted production nearly 30 percent but cut energy use 7 percent and greenhouse gas emissions 72 percent (measured in terms of their carbon dioxide equivalent), saving more than $2 billion so far. Five other major firms—IBM, British Telecom, Alcan, NorskeCanada and Bayer—have collectively saved at least another $2 billion since the early 1990s by reducing their carbon emissions more than 60 percent. In 2001 oil giant BP met its 2010 goal of reducing carbon dioxide emissions 10 percent below the company's 1990 level, thereby cutting its energy bills $650 million over 10 years. And just this past May, General Electric vowed to raise its energy efficiency 30 percent by 2012 to enhance the company's shareholder value. These sharp-penciled firms, and dozens like them, know that energy efficiency improves the bottom line and yields even more valuable side benefits: higher quality and reliability in energy-efficient factories, 6 to 16 percent higher labor productivity in efficient offices, and 40 percent higher sales in stores skillfully designed to be illuminated primarily by daylight.

The U.S. now uses 47 percent less energy per dollar of economic output than it did 30 years ago, lowering costs by $1 billion a day. These savings act like a huge universal tax cut that also reduces the federal deficit. Far from dampening global development, lower energy bills accelerate it. And there is plenty more value to capture at every stage of energy production, distribution and consumption. Converting coal at the power plant into incandescent light in your house is only 3 percent efficient. Most of the waste heat discarded at

Overview: Crossroads for Energy

THE PROBLEM

The energy sector of the global economy is woefully inefficient. Power plants and buildings waste huge amounts of heat, cars and trucks dissipate most of their fuel energy and consumer appliances waste much of their power (and often siphon electricity even when they are turned off).

If nothing is done, the use of oil and coal will continue to climb, draining hundreds of billions of dollars a year from the economy as well as worsening the climate, pollution and oil-security problems.

THE PLAN

Improving end-use efficiency is the fastest and most lucrative way to save energy. Many energy-efficient products cost no more than inefficient ones. Homes and factories that use less power can be cheaper to build than conventional structures. Reducing the weight of vehicles can double their fuel economy without compromising safety or raising sticker prices.

With the help of efficiency improvements and competitive renewable energy sources, the U.S. can phase out oil use by 2050. Profit-seeking businesses can lead the way.

U.S. power stations—which amounts to 20 percent more energy than Japan uses for everything—could be lucratively recycled. About 5 percent of household electricity in the U.S. is lost to

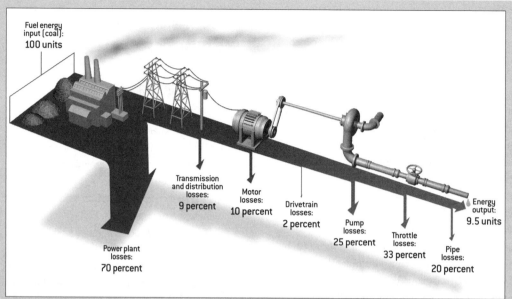

Fuel energy input (coal): **100 units**

Transmission and distribution losses: **9 percent**

Motor losses: **10 percent**

Drivetrain losses: **2 percent**

Pump losses: **25 percent**

Throttle losses: **33 percent**

Pipe losses: **20 percent**

Energy output: **9.5 units**

Power plant losses: **70 percent**

DON FOLEY

Compounding Losses

From the power plant to an industrial pipe, inefficiencies along the way whittle the energy input of the fuel—set at 100 arbitrary units in this example—by more than 90 percent, leaving only 9.5 units of energy delivered as fluid flow through the pipe. But small increases in end-use efficiency can reverse these compounding losses. For instance, saving one unit of output energy by reducing friction inside the pipe will cut the needed fuel input by 10 units, slashing cost and pollution at the power plant while allowing the use of smaller, cheaper pumps and motors.

energizing computers, televisions and other appliances that are turned off. (The electricity wasted by poorly designed standby circuitry is equivalent to the output of more than a dozen 1,000-megawatt power stations running full tilt.) In all, preventable energy waste costs Americans hundreds of billions of dollars and the global economy more than $1 trillion a year, destabilizing the climate while producing no value.

If energy efficiency has so much potential, why isn't everyone pursuing it? One obstacle is that many people have confused efficiency (doing more with less) with curtailment, discomfort or privation (doing less, worse or without). Another obstacle is that energy users do not recognize how much they can benefit from improving efficiency, because saved energy comes in millions of invisibly small pieces, not in obvious big chunks. Most people lack the time and attention to learn about modern efficiency techniques, which evolve so quickly that even experts cannot keep up. Moreover, taxpayer-funded subsidies have made energy seem cheap. Although the U.S. government has declared that bolstering efficiency is a priority, this commitment is mostly rhetorical. And scores of ingrained rules and habits block efficiency efforts or actually reward waste. Yet relatively simple changes can turn all these obstacles into business opportunities.

Enhancing efficiency is the most vital step toward creating a climate-safe energy system, but switching to fuels that emit less carbon will also play an important role. The world economy is already decarbonizing: over the past two centuries, carbon-rich fuels such as coal have given way to fuels with less carbon (oil and natural gas) or with none (renewable sources such as solar and wind power). Today less than one-third of the fossil fuel atoms burned are carbon; the rest are climate-safe hydrogen. This decarbonization trend is reinforced by greater efficiencies in converting, distributing and using energy; for example, combining the production of heat and electricity can extract twice as much useful work from each ton of carbon emitted into the atmosphere. Together these advances could dramatically reduce total carbon emissions by 2050 even as the global economy expands. This article focuses on the biggest prize: wringing more work from each unit of energy delivered to businesses and consumers. Increasing end-use efficiency can yield huge savings in fuel, pollution and capital costs because large amounts of energy are lost at every stage of the journey from production sites to delivered services [see sidebar, "Compounding Losses," page 170]. So even small reductions in the power used at the downstream end of the chain can enormously lower the required input at the upstream end.

The Efficiency Revolution

Many energy-efficient products, once costly and exotic, are now inexpensive and commonplace. Electronic speed controls, for example, are mass-produced so cheaply that some suppliers give them away as a free bonus with each motor. Compact fluorescent lamps cost more than $20 two decades ago but only $2 to $5 today; they use 75 to 80 percent less electricity than incandescent bulbs and last 10 to 13 times longer. Window coatings that transmit light but reflect heat cost one-fourth of what they did five years ago. Indeed, for many kinds of equipment in competitive markets—motors, industrial pumps, televisions, refrigerators—some highly energy-efficient models cost no more than inefficient ones. Yet far more important than all these better and cheaper technologies is a hidden revolution in the design that combines and applies them.

For instance, how much thermal insulation is appropriate for a house in a cold climate? Most engineers would stop adding insulation when the expense of putting in more material rises above the savings over time from lower heating bills. But this comparison omits the capital cost of the heating system—the furnace, pipes, pumps, fans and so on—which may not be necessary at all if the insulation is good enough. Consider my own house, built in 1984 in Snowmass, Colorado, where winter temperatures can dip to –44 degrees Celsius and frost can occur any day of the year. The house has no conventional heating system; instead, its roof is insulated with 20 to 30 centimeters of polyurethane foam, and its 40-centimeter-thick masonry walls sandwich another 10 centimeters of the material. The double-pane windows combine two or three transparent heat-reflecting films with insulating krypton gas, so that they block heat as well as eight to 14 panes of glass. These features, along with heat recovery from the ventilated air, cut the house's heat losses to only about 1 percent more than the heat gained from sunlight, appliances and people inside the structure. I can offset this tiny loss by playing with my dog (who generates about 50 watts of heat, adjustable to 100

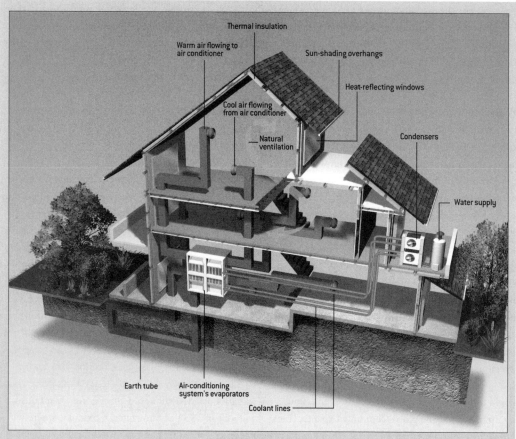

Thermal insulation

Warm air flowing to
air conditioner

Sun-shading overhangs

Heat-reflecting windows

Cool air flowing
from air conditioner

Natural
ventilation

Condensers

Water supply

Earth tube

Air-conditioning
system's evaporators

Coolant lines

DON FOLEY

Saving Energy by Design

How can you keep cool in tropical Thailand while minimizing power usage? Architect Soontorn Boonyatikarn of Chulalongkorn University used overhangs and balconies to shade his 350-square-meter home in Pathumthani, near Bangkok. Insulation, an airtight shell and infrared-reflecting windows keep heat out of the house while letting in plenty of daylight.

An open floor plan and central stairwell promote ventilation, and indoor air is cooled as it flows through an underground tube. As a result, the house needs just one-seventh of the typical air-conditioning capacity for a structure of its size. To further reduce energy bills, the air-conditioning system's condensers heat the house's water.

watts if you throw a ball to her) or by burning obsolete energy studies in a small woodstove on the coldest nights.

Eliminating the need for a heating system reduced construction costs by $1,100 (in 1983 dollars). I then reinvested this money, plus another $4,800, into equipment that saved half the water, 99 percent of the water-heating energy and 90 percent of the household electricity. The 4,000-square-foot structure—which also houses the original headquarters of Rocky Mountain Institute (RMI), the nonprofit group I cofounded in 1982—consumes barely more electricity than a single 100-watt lightbulb. (This amount excludes the power used by the institute's office equipment.) Solar cells generate five to six times that much electricity, which I sell back to the utility. Together all the efficiency investments repaid their cost in 10 months with 1983 technologies; today's are better and cheaper.

In the 1990s Pacific Gas & Electric undertook an experiment called ACT^2 that applied smart design in seven new and old buildings to demonstrate that large efficiency improvements can be cheaper than small ones. For example, the company built a new suburban tract house in Davis, California, that could stay cool in the summer without air-conditioning. PG&E estimated that such a design, if widely adopted, would cost about $1,800 less to build and $1,600 less to maintain over its lifetime than a conventional home of the same size. Similarly, in 1996 Thai architect Soontorn Boonyatikarn built a house near steamy Bangkok that required only one-seventh the air-conditioning capacity usually installed in a structure of that size; the savings in equipment costs paid for the insulating roof, walls and windows that keep the house cool [see sidebar, "Saving Energy by Design," page 172]. In all these cases, the design approach was the same: optimize the whole building for multiple benefits rather than use isolated components for single benefits.

Such whole-system engineering can also be applied to office buildings and factories. The designers of a carpet factory built in Shanghai in 1997 cut the pumping power required for a heat-circulating loop by 92 percent through two simple changes. The first change was to install fat pipes rather than thin ones, which greatly reduced friction and hence allowed the system to use smaller pumps and motors. The second innovation was to lay out the pipes before positioning the equipment they connect. As a result, the fluid moved through short, straight pipes instead of tracing circuitous paths, further reducing friction and capital costs.

This isn't rocket science; it's just good Victorian engineering rediscovered. And it is widely applicable. A practice team at RMI has recently developed new-construction designs offering energy savings of 89 percent for a data center, about 75 percent for a chemical plant, 70 to 90 percent for a supermarket and about 50 percent for a luxury yacht, all with capital costs lower than those of conventional designs. The team has also proposed retrofits for existing oil refineries, mines and microchip factories that would reduce energy use by 40 to 60 percent, repaying their cost in just a few years.

Vehicles of Opportunity

Transportation consumes 70 percent of U.S. oil and generates a third of the nation's carbon emissions. It is widely considered the most intractable part of the climate problem, especially as hundreds of millions of people in China and India buy automobiles. Yet transportation offers enormous efficiency opportunities. *Winning the Oil Endgame,* a 2004 analysis written by my team at RMI and cosponsored by the Pentagon, found that artfully combining lightweight materials with innovations in propulsion and aerodynamics could cut oil use by cars, trucks and planes by two-thirds without compromising comfort, safety, performance or affordability.

Despite 119 years of refinement, the modern car remains astonishingly inefficient. Only 13 percent of its fuel energy even reaches the wheels—the other 87 percent is either dissipated as heat and noise in the engine and drivetrain or lost to idling and accessories such as air conditioners. Of the energy delivered to the wheels, more than half heats the tires, road and air. Just 6 percent of the fuel energy actually accelerates the car (and all this energy converts to brake heating when you stop). And because 95 percent of the accelerated mass is the car itself, less than 1 percent of the fuel ends up moving the driver.

Yet the solution is obvious from the physics: greatly reduce the car's weight, which causes three-fourths of the energy losses at the wheels. And every unit of energy saved at the wheels by lowering weight (or cutting drag) will save an additional seven units of energy now lost en route to the wheels. Concerns about cost and safety have long discouraged attempts to make lighter cars, but modern light-but-strong materials—new metal alloys and advanced polymer composites—can slash a car's mass without sacrificing crashworthiness. For example, carbon-fiber composites can absorb six to 12 times as much crash energy per kilogram as steel does, more than offsetting the composite car's weight disadvantage if it hits a steel vehicle that is twice as heavy. With such novel materials, cars can be big, comfortable and protective without being heavy, inefficient and hostile, saving both oil *and* lives. As Henry Ford said, you don't need weight for strength; if you did, your bicycle helmet would be made of steel, not carbon fiber.

Advanced manufacturing techniques developed in the past two years could make carbon-composite car bodies competitive with steel ones. A lighter body would allow automakers to use smaller (and less expensive) engines. And because the assembly of carbon-composite cars does not require body or paint shops, the factories would be smaller and

cost 40 percent less to build than conventional auto plants. These savings would offset the higher cost of the carbon-composite materials. In all, the introduction of ultralight bodies could nearly double the fuel efficiency of today's hybrid-electric vehicles—which are already twice as efficient as conventional cars—without raising their sticker prices. If composites prove unready, new ultralight steels offer a reliable backstop. The competitive marketplace will sort out the winning materials, but either way, superefficient ultralight vehicles will start pulling away from the automotive pack within the next decade.

What is more, ultralight cars could greatly accelerate the transition to hydrogen fuel-cell cars that use no oil at all [see "On the Road to Fuel-Cell Cars," by Steven Ashley; *Scientific American,* March 2005]. A midsize SUV whose halved weight and drag cut its needed power to the wheels by two-thirds would have a fuel economy equivalent to 114 miles per gallon and thus require only a 35-kilowatt fuel cell—one-third the usual size and hence much easier to manufacture affordably [see sidebar, "A Lean Mean Driving Machine," page 175]. And because the vehicle would need to carry only one-third as much hydrogen, it would not require any new storage technologies; compact, safe, off-the-shelf carbon-fiber tanks could hold enough hydrogen to propel the SUV for 530 kilometers. Thus, the first automaker to go ultralight will win the race to fuel cells, giving the whole industry a strong incentive to become as boldly innovative in materials and manufacturing as a few companies now are in propulsion.

RMI's analysis shows that full adoption of efficient vehicles, buildings and industries could shrink projected U.S. oil use in 2025—28 million barrels a day—by more than half, lowering consumption to pre-1970 levels. In a realistic scenario, only about half of these savings could actually be captured by 2025 because many older, less efficient

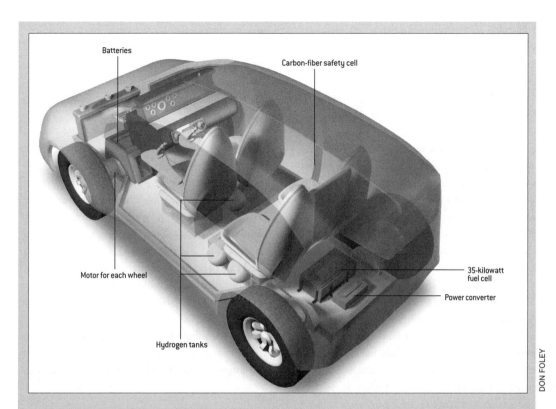

Batteries

Carbon-fiber safety cell

Motor for each wheel

35-kilowatt fuel cell

Power converter

Hydrogen tanks

DON FOLEY

A Lean Mean Driving Machine

Ultralight cars can be fast, roomy, safe and efficient. A concept five-seat mid-size SUV called the Revolution, designed in 2000, weighs only 857 kilograms—less than half the weight of a comparable conventional car—yet its carbon-fiber safety cell would protect passengers from high-speed collisions with much heavier vehicles. A 35-kilowatt fuel cell could propel the car for 530 kilometers on 3.4 kilograms of hydrogen stored in its tanks. And the Revolution could accelerate to 100 kilometers per hour in 8.3 seconds.

cars and trucks would remain on the road (vehicle stocks turn over slowly). Before 2050, though, U.S. oil consumption could be phased out altogether by doubling the efficiency of oil use and substituting alternative fuel supplies. Businesses can profit greatly by making the transition, because saving each barrel of oil through efficiency improvements costs only $12, less than one-fifth of what petroleum sells for today. And two kinds of alternative fuel supplies could compete robustly with oil even if it sold for less than half the current price. The first is ethanol made from woody, weedy plants such as switchgrass and poplar. Corn is currently the main U.S. source of ethanol, which is blended with gasoline, but the woody plants yield twice as much ethanol per ton as corn does and with lower capital investment and far less energy input.

The second alternative is replacing oil with lower-carbon natural gas, which would become cheaper and more abundant as efficiency gains reduce the demand for electricity at peak periods. At those times, gas-fired turbines generate power so wastefully that saving 1 percent of electricity would cut U.S. natural gas consumption by 2 percent and its price by 3 or 4 percent. Gas saved in this way and in other uses could then replace oil either directly or, even more profitably and efficiently, by converting it to hydrogen.

The benefits of phasing out oil would go far beyond the estimated $70 billion saved every year. The transition would lower U.S. carbon emissions by 26 percent and eliminate all the social and political costs of getting and burning petroleum—military conflict, price volatility, fiscal and diplomatic distortions, pollution and so on. If the country becomes oil free, then petroleum will no longer be worth fighting over. The Pentagon would also reap immediate rewards from raising energy efficiency because it badly needs to reduce the costs and risks of supplying fuel to its troops. Just as the U.S. Department of Defense's research efforts transformed civilian industry by creating the Internet and the Global Positioning System, it should now spearhead the development of advanced ultralight materials.

The switch to an oil-free economy would happen even faster than RMI projected if policymakers stopped encouraging the perverse development patterns that make people drive so much. If federal, state and local governments did not mandate and subsidize suburban sprawl, more of us could live in neighborhoods where almost everything we want is within a five-minute walk. Besides saving fuel, this New Urbanist design builds stronger communities, earns more money for developers and is much less disruptive than other methods of limiting vehicle traffic (such as the draconian fuel and car taxes that Singapore uses to avoid Bangkok-like traffic jams).

Renewable Energy

Efficiency improvements that can save most of our electricity also cost less than what the utilities now pay for coal, which generates half of U.S. power and 38 percent of its fossil fuel carbon emissions. Furthermore, in recent years alternatives to coal-fired power plants—including renewable sources such as wind and solar power, as well as decentralized cogeneration plants that produce electricity and heat together in buildings and factories—have begun to hit their stride. Worldwide, the collective generating capacity of these sources is already greater than that of nuclear power and growing six times as fast [see illustration, "Electricity Alternatives," below]. This trend is all the more

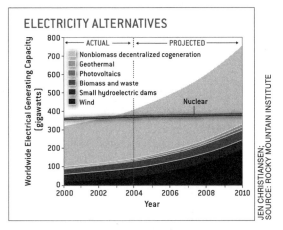

Decentralized sources of electricity—cogeneration (the combined production of electricity and heat, typically from natural gas) and renewables (such as solar and wind power)—surpassed nuclear power in global generating capacity in 2002. The annual output of these low- and no-carbon sources will exceed that of nuclear power this year.

impressive because decentralized generators face many obstacles to fair competition and usually get much lower subsidies than centralized coal-fired or nuclear plants.

Wind power is perhaps the greatest success story. Mass production and improved engineering have made modern wind turbines big (generating two to five megawatts each), extremely reliable and environmentally quite benign. Denmark already gets a fifth of its electricity from wind, Germany a tenth. Germany and Spain are each adding more than 2,000 megawatts of wind power each year, and Europe aims to get 22 percent of its electricity and 12 percent of its total energy from renewables by 2010. In contrast, global nuclear generating capacity is expected to remain flat, then decline.

The most common criticism of wind power—that it produces electricity too intermittently—has not turned out to be a serious drawback. In parts of Europe that get all their power from wind on some days, utilities have overcome the problem by diversifying the locations of their wind turbines, incorporating wind forecasts into their generating plans and integrating wind power with hydroelectricity and other energy sources. Wind and solar power work particularly well together, partly because the conditions that are bad for wind (calm, sunny weather) are good for solar and vice versa. In fact, when properly combined, wind and solar facilities are more reliable than conventional power stations—they come in smaller modules (wind turbines, solar cells) that are less likely to fail all at once, their costs do not swing wildly with the prices of fossil fuels, and terrorists are much more likely to attack a nuclear reactor or an oil terminal than a wind farm or a solar array.

Most important, renewable power now has advantageous economics. In 2003 U.S. wind energy sold for as little as 2.9 cents a kilowatt-hour. The federal government subsidizes wind power with a production tax credit, but even with-out that subsidy, the price—about 4.6 cents per kilowatt-hour—is still cheaper than subsidized power from new coal or nuclear plants. (Wind power's subsidy is a temporary one that Congress has repeatedly allowed to expire; in contrast, the subsidies for the fossil fuel and nuclear industries are larger and permanent.) Wind power is also abundant: wind farms occupying just a few percent of the available land in the Dakotas could cost-effectively meet all of America's electricity needs. Although solar cells currently cost more per kilowatt-hour than wind turbines do, they can still be profitable if integrated into buildings, saving the cost of roofing materials. Atop big, flat-roofed commercial buildings, solar cells can compete without subsidies if combined with efficient use that allows the building's owner to resell the surplus power when it is most plentiful and valuable—on sunny afternoons. Solar is also usually the cheapest way to get electricity to the two billion people, mostly in the developing world, who have no access to power lines. But even in rich countries, a house as efficient as mine can get all its electricity from just a few square meters of solar cells, and installing the array costs less than connecting to nearby utility lines.

Cheaper to Fix

Inexpensive efficiency improvements and competitive renewable sources can reverse the terrible arithmetic of climate change, which accelerates exponentially as we burn fossil fuels ever faster. Efficiency can outpace economic growth if we pay attention: between 1977 and 1985, for example, U.S. gross domestic product (GDP) grew 27 percent, whereas oil use fell 17 percent. (Over the same period, oil imports dropped 50 percent, and Persian Gulf imports plummeted 87 percent.) The growth of renewables has routinely outpaced GDP; worldwide, solar and wind power are doubling every two and three years, respectively. If both efficiency and renewables grow faster than the economy, then carbon emissions will fall and global warming will slow—buying more time to develop even better technologies for displacing the remaining fossil fuel use or to master and deploy ways to capture combustion carbon before it enters the air [see "Can We Bury Global Warming?" page 44].

In contrast, nuclear power is a slower and much more expensive solution. Delivering a kilowatt-hour from a new nuclear plant costs at least three times as much as saving one through efficiency measures. Thus, every dollar spent on efficiency would displace at least three times as much coal as spending on nuclear power, and the efficiency improvements could go into effect much more quickly because it takes so long to build reactors. Diverting public and private investment from market winners to losers does not just distort markets and misallocate financial capital—it worsens the climate problem by buying a less effective solution.

The good news about global warming is that it is cheaper to fix than to ignore. Because saving energy is profitable, efficient use is gaining traction in the marketplace. U.S. Environmental Protection Agency economist Skip Laitner calculates that from 1996 to mid-2005 prudent choices by businesses and consumers, combined with the shift to a more information- and service-based economy, cut average U.S. energy use per dollar of GDP by 2.1 percent a year—nearly three times as fast as the rate for the preceding 10 years. This change met 78 percent of the rise in demand for energy services over the past decade (the remainder was met by increasing energy supply), and the U.S. achieved this progress without the help of any technological breakthroughs or new national policies. The climate problem was created by millions of bad decisions over decades, but climate stability can be restored by millions of sensible choices—buying a more efficient lamp or car, adding insulation or caulk to your home, repealing subsidies for

waste and rewarding desired outcomes (for example, by paying architects and engineers for savings, not expenditures).

The proper role of government is to steer, not row, but for years officials have been steering our energy ship in the wrong direction. The current U.S. energy policy harms the economy and the climate by rejecting free-market principles and playing favorites with technologies. The best course is to allow every method of producing or saving energy to compete fairly, at honest prices, regardless of which kind of investment it is, what technology it uses, how big it is or who owns it. For example, few jurisdictions currently let decentralized power sources such as rooftop solar arrays "plug and play" on the electric grid, as modern technical standards safely permit. Although 31 U.S. states allow net metering—the utility buys your power at the same price it charges you—most artificially restrict or distort this competition. But the biggest single obstacle to electric and gas efficiency is that most countries, and all U.S. states except California and Oregon, reward distribution utilities for selling more energy and penalize them for cutting their customers' bills. Luckily, this problem is easy to fix: state regulators should align incentives by decoupling profits from energy sales, then letting utilities keep some of the savings from trimming energy bills.

Superefficient vehicles have been slow to emerge from Detroit, where neither balance sheets nor leadership has supported visionary innovation. Also, the U.S. lightly taxes gasoline but heavily subsidizes its production, making it cheaper than bottled water. Increasing fuel taxes may not be the best solution, though; in Europe, stiff taxes—which raise many countries' gasoline prices to $4 or $5 a gallon—cut driving more than they make new cars efficient, because fuel costs are diluted by car owners' other expenses and are then steeply discounted (most car buyers count only the first few years' worth of fuel savings). Federal standards adopted in the 1970s helped to lift the fuel economy of new cars and light trucks from 16 miles per gallon in 1978 to 22 miles per gallon in 1987, but the average has slipped to 21 mpg since then. The government projects that the auto industry will spend the next 20 years getting its vehicles to be just 0.5 mile per gallon more efficient than they were in 1987. Furthermore, automakers loathe the standards as restrictions on choice and have become adept at gaming the system by selling more vehicles classified as light trucks, which are allowed to have lower fuel economy than cars. (The least efficient light trucks even get special subsidies.)

The most powerful policy response is "feebates"—charging fees on inefficient new cars and returning that revenue as rebates to buyers of efficient models. If done separately for each size class of vehicle, so there is no bias against bigger models, feebates would expand customer choice instead of restricting it. Feebates would also encourage innovation, save customers money and boost automakers' profits. Such policies, which can be implemented at the state level, could speed the adoption of advanced-technology cars, trucks and planes without mandates, taxes, subsidies or new national laws.

In Europe and Japan, the main obstacle to saving energy is the mistaken belief that their economies are already as efficient as they can get. These countries are up to twice as efficient as the U.S., but they still have a long way to go. The greatest opportunities, though, are in developing countries, which are on average three times less efficient than the U.S. Dreadfully wasteful motors, lighting ballasts and other devices are freely traded and widely bought in these nations. Their power sector currently devours one-quarter of their development funds, diverting money from other vital projects. Industrial countries are partly responsible for this situation because many have exported inefficient vehicles and equipment to the developing world. Exporting inefficiency is both immoral and uneconomic; instead, the richer nations should help developing countries build an energy-efficient infrastructure that would free up capital to address their other pressing needs. For example, manufacturing efficient lamps and windows takes 1,000 times less capital than building power plants and grids to do the same tasks, and the investment is recovered 10 times faster.

China and India have already discovered that their burgeoning economies cannot long compete if energy waste continues to squander their money, talent and public health. China is setting ambitious but achievable goals for shifting from coal-fired power to decentralized renewable energy and natural gas. (The Chinese have large supplies of gas and are expected to tap vast reserves in eastern Siberia.) Moreover, in 2004 China announced an energy strategy built around "leapfrog technologies" and rapid improvements in the efficiency of new buildings, factories and consumer products. China is also taking steps to control the explosive growth of its oil use; by 2008 it will be illegal to sell many inefficient U.S. cars there. If American automakers do not innovate quickly enough, in another decade you may well be driving a superefficient Chinese-made car. A million U.S. jobs hang in the balance.

Today's increasingly competitive global economy is stimulating an exciting new pattern of energy investment. If governments can remove institutional barriers and harness the dynamism of free enterprise, the markets will naturally favor choices that generate wealth, protect the climate and build real security by replacing fossil fuels with cheaper alternatives. This technology-driven convergence of business, environmental and security interests—creating abundance by design—holds out the promise of a fairer, richer and safer world.

—SEPTEMBER 2005

An Efficient Solution

*Wasting less energy is the quickest,
least expensive way to stem carbon emissions.*

EBERHARD K. JOCHEM

The huge potential of energy-efficiency measures for mitigating the release of greenhouse gases into the atmosphere attracts little attention when placed alongside the more glamorous alternatives of nuclear, hydrogen or renewable energies. But developing a comprehensive efficiency strategy is the fastest and cheapest thing we can do to reduce carbon emissions. It can also be profitable and astonishingly effective, as two recent examples demonstrate.

From 2001 through 2005, Procter & Gamble's factory in Germany increased production by 45 percent, but the energy needed to run machines

and to heat, cool and ventilate buildings rose by only 12 percent, and carbon emissions remained at the 2001 level. The major pillars supporting this success include highly efficient illumination, compressed-air systems, new designs for heating and air-conditioning, funneling heat losses from compressors into heating buildings, and detailed energy measurement and billing.

In some 4,000 houses and buildings in Germany, Switzerland, Austria and Scandinavia, extensive insulation, highly efficient windows and energy-conscious design have led to enormous efficiency increases, enabling energy budgets for heating that are a sixth of the requirement for typical buildings in these countries.

Improved efficiencies can be realized all along the energy chain, from the conversion of primary energy (oil, for example) to energy carriers (such as electricity) and finally to useful energy (the heat in your toaster). The annual global primary energy demand is 447,000 petajoules (a petajoule is roughly 300 gigawatt-hours), 80 percent of which comes from carbon-emitting fossil fuels such as coal, oil and gas. After conversion these primary energy sources deliver roughly 300,000 petajoules of so-called final energy to customers in the form of electricity, gasoline, heating oil, jet fuel, and so on.

The next step, the conversion of electricity, gasoline, and the like to useful energy in engines,

Overview

- Two-thirds of all energy is lost during its conversion into forms used in human activities; most of this energy comes from carbon-emitting fossil fuels.

- The quickest, easiest way to reduce carbon emissions is to avoid as many of these losses as possible.

- Improving the energy efficiency of buildings, appliances and industrial processes offers impressive savings.

boilers and lightbulbs, causes further energy losses of 154,000 petajoules. Thus, at present almost 300,000 petajoules, or two-thirds of the primary energy, are lost during the two stages of energy conversion. Furthermore, all useful energy is eventually dissipated as heat at various temperatures. Insulating buildings more effectively, changing industrial processes and driving lighter, more aerodynamic cars [see "Fueling Our Transportation Future," page 162] would reduce the demand for useful energy, thus substantially reducing energy wastage.

Given the challenges presented by climate change and the high increases expected in energy prices, the losses that occur all along the energy chain can also be viewed as opportunities—and efficiency is one of the most important. New technologies and know-how must replace the present intensive use of energy and materials.

Room for Improvement

Because conservation measures, whether incorporated into next year's car design or a new type of power plant, can have a dramatic impact on energy consumption, they also have an enormous effect on overall carbon emissions. In this mix, buildings and houses, which are notoriously inefficient in many countries today, offer the greatest potential for saving energy. In countries belonging to the Organization for Economic Cooperation and Development (OECD) and in the megacities of emerging countries, buildings contribute more than one-third of total energy-related greenhouse gas emissions.

Little heralded but impressive advances have already been made, often in the form of efficiency improvements that are invisible to the consumer. Beginning with the energy crisis in the 1970s, air conditioners in the U.S. were redesigned to use less power with little loss in cooling capacity, and new U.S. building codes required more insulation and double-paned windows. New refrigerators use only one quarter of the power of earlier models. (With approximately 150 million refrigerators and freezers in the U.S., the difference in consumption between 1974 efficiency levels and 2001 levels is equivalent to avoiding the generation of 40 gigawatts at power plants.) Changing to compact fluorescent lightbulbs yields an instant reduction in power demand; these bulbs provide as much light as regular incandescent bulbs, last 10 times longer and use just one-fourth to one-fifth the energy.

Despite these gains, the biggest steps remain to be taken. Many buildings were designed with the intention of minimizing construction costs rather than life-cycle cost, including energy use, or simply in ignorance of energy-saving considerations. Take roof overhangs, for example, which in warm climates traditionally measured a meter or so and which are rarely used today because of the added cost, although they would control heat buildup on walls and windows. One of the largest European manufacturers of prefabricated houses is now offering zero-net-energy houses: these well-insulated and intelligently designed structures with solar-thermal and photovoltaic collectors do not need commercial energy, and their total cost is similar to those of new houses built to conform to current building codes. Because buildings have a 50- to 100-year lifetime, efficiency retrofits are essential. But we need to coordinate changes in existing buildings thoughtfully to avoid replacing a single component, such as a furnace, while leaving in place leaky ducts and single-pane windows that waste much of the heat the new furnace produces.

One example highlights what might be done in industry: although some carpet manufacturers still dye their products at 100 to 140 degrees Celsius, others dye at room temperature using enzyme technology, reducing the energy demand by more than 90 percent.

The Importance of Policy

To realize the full benefits of efficiency, strong energy policies are essential. Among the underlying reasons for the crucial role of policy are the dearth of knowledge by manufacturers and the public about efficiency options, budgeting methods that do not take proper account of the ongoing benefits of long-lasting investments, and market imperfections such as external costs for carbon emissions and other costs of energy use. Energy policy set by governments has traditionally underestimated the benefits of efficiency. Of course, factors other than policy can drive changes in efficiency—higher energy prices, new technologies or cost competition, for instance. But policies—which include energy taxes, financial incentives, professional training, labeling, environmental legislation, greenhouse gas emissions trading and international coordination of regulations for traded products—can make an enormous difference. Furthermore, rapid growth in demand for energy services in emerging countries provides an opportunity to implement energy-efficient policies from the outset as infrastructure grows: programs to realize efficient solutions in buildings, transport systems and industry would give people the energy services they need without having to build as many power plants, refineries or gas pipelines.

Japan and the countries of the European Union have been more eager to reduce oil imports than the U.S. has and have encouraged productivity gains through energy taxes and other measures. But all OECD countries except Japan have so far failed to update appliance standards. Nor do gas and electric bills in OECD countries indicate how much energy is used for heating, say, as opposed to boiling water or which uses are the most energy intensive—that is, where a reduction in usage would produce the greatest energy savings. In industry, compressed air, heat, cooling and electricity are often not billed by production line but expressed as an overhead cost.

Nevertheless, energy efficiency has a higher profile in Europe and Japan. A retrofitting project in Ludwigshafen, Germany, serves as just one example. Five years ago 500 dwellings were equipped to adhere to low-energy standards (about 30 kilowatt-hours per square meter per year), reducing the annual energy demand for heating those buildings by a factor of six. Before the retrofit, the dwellings were difficult to rent; now demand is three times greater than capacity.

Other similar projects abound. The Board of the Swiss Federal Institutes of Technology, for instance, has suggested a technological program aimed at what we call the 2,000-Watt Society—an annual primary energy use of 2,000 watts (or 65 gigajoules) per capita. Realizing this vision in industrial countries would reduce the per capita energy use and related carbon emissions by two-thirds, despite a two-thirds increase in GDP, within the next 60 to 80 years. Swiss scientists, including myself, have been evaluating this plan since 2002, and we have concluded that the goal of the 2,000-watt-per-capita society is technically feasible for industrial countries in the second half of this century.

To some people, the term "energy efficiency" implies reduced comfort. But the concept of efficiency means that you get the same service—a comfortable room or convenient travel from home to work—using less energy. The EU, its member states and Japan have begun to tap the substantial—and profitable—potential of efficiency measures. To avoid the rising costs of energy supplies and the even costlier adaptations to climate change, efficiency must become a global activity.

—SEPTEMBER 2006

In Pursuit of the Ultimate Lamp

Full-spectrum light-emitting diodes, or LEDs, are becoming widespread—and the race is on to develop white-light versions to replace Edison's century-old incandescent bulb.

M. GEORGE CRAFORD, NICK HOLONYAK, JR., AND FREDERICK A. KISH, JR.

In 1995 one of us (Holonyak) was honored to accept the Japan Prize for pioneering work in semiconductor light emitters and lasers. Asked to say a few words about tomorrow's technology, he simply pointed to the ceiling lights and said, "All of this is going."

A revolution is taking place, literally in front of our eyes, thanks to semiconductor devices known as light-emitting diodes, or LEDs. Most familiar as the little glowing red or green indicator lights on electronic equipment, LEDs are beginning to replace incandescent bulbs in many applications. The reason? LEDs convert electricity to colored light more efficiently than their incandescent cousins—for red light, their efficiency is 10 times greater. They are rugged and compact; some types last a phenomenal 100,000 hours, or about a decade of regular use. In contrast, the average incandescent bulb lasts about 1,000 hours. Moreover, the intensity and colors of LED light have improved so much that the diodes are now suitable for large displays—perhaps the most impressive example being the eight-story-tall Nasdaq billboard in New York City's Times Square.

Currently, engineers are trying to lower the cost of manufacturing LEDs, improve their efficiency and extend their range of useful colors. In fact, it is possible to combine the output of red, green and blue LEDs to make white light, a cheap, mass-market form of which would be the brilliant prize of the industry. Such an LED could someday supplant, more than a century after Thomas Edison's invention, the incandescent lightbulb.

LEDs are already replacing lightbulbs in several instances, albeit in a fashion less dramatic than the giant Nasdaq display. The new applications are perhaps most noticeable to automobile drivers. In Europe 60 to 70 percent of the cars produced use LEDs for their high-mount brake lights, and the U.S. is beginning to move in that direction, too. LEDs are also being used for taillights and turn signals, as well as for side markers for trucks and buses. We expect that by the end of the decade LEDs will dominate the red and amber lighting on the exterior of vehicles. Larger and brighter LEDs are making their way into the red of traffic lights. About 10 percent of the nation's stoplights include LEDs.

Traditionally, traffic signals and other colored lamps use incandescent bulbs, which are then covered with a filter to produce the appropriate hue. Although filtering is cheap—the bulbs emit light that costs a mere fraction of a cent per lumen (the standard measurement unit of illumination)—it is a terribly inefficient way to produce light. A red filter, for example, blocks about 80 percent of the glow, so the amount of light that emerges drops from about 17 lumens per watt of power to three to five lumens per watt.

In contrast, lumens cast by an LED stoplight may cost around 15 cents each to produce, but virtually all of them are of the right color. What is more, the LEDs in a stoplight consume only 10 to 25 watts, compared with the 50 to 150 watts used by an incandescent bulb of similar brightness. This energy savings pays for the higher cost of an LED in as little as one year. When this figure is considered with the reduced maintenance and labor costs of LEDs, it is easy to see why LEDs are becoming more popular with city planners.

Interior designers began to use LEDs a few years ago, when high-brightness models of all colors made their appearance. Because each LED gives off one distinct hue, users can have complete control of nearly the full spectrum. By putting differently colored LEDs together in an array, the user can adjust their combined light. For example, white light composed of red, green and blue LEDs can be made to feel "cooler" by turning off more of the red LEDs and turning on more of the blue ones.

This flexibility offers novel ways of using light. Rather than, say, putting up new wallpaper or applying fresh paint, one can manipulate the color of a room by adjusting the ratio of wavelengths in the emitted light (the wavelengths of light determine the colors). The Metropolitan Museum of Art in New York City relied on just this kind of LED lighting to illuminate its display of the Beatles' *Sgt. Pepper's* costumes in 1999. (The lighting is also cool and will not damage the fabric.) Photographers would have total control of artificial light sources without the need for cumbersome filters or gels.

LEDs also offer intriguing possibilities for medical science. For example, LEDs' cool temperatures, precise wavelength control and broad-beam characteristics enable cancer researchers to study the photodynamic treatment of tumors more effectively. In this therapy, patients receive light-sensitive drugs that are preferentially absorbed by tumor cells. When light of the appropriate wavelength strikes these chemicals, they become excited and destroy the cells. By using an array of LEDs linked together, researchers can produce an even "sheet" of light that stimulates light-sensitive drugs without burning the patient's skin.

Up Close

Maybe the easiest way to examine light-emitting diodes in detail is to purchase a few of them, in the form of a $15 flashing-red LED bicycle light. Open up the casing, and you'll see a pair of AA batteries wired to a circuit board containing a series of clear, colorless, cylindrical knobs approximately five-sixteenths of an inch high and three-sixteenths of an inch in diameter. Each of these transparent knobs is a light-emitting diode. Press the "on" but-

SEE *Figure 11 in color section.*

ton, and the clear LED turns red, casting a color so brilliant that it can be painful to look at directly. If you turn it off and closely examine the LED, you will see what is the equivalent of a wire threading through its base—and what looks like a miniature cup about halfway up. This cup is a reflector, which holds a semiconductor chip about the size of a grain of sand. This chip is the LED's "heart."

Inside the chip, there is a layer that has an excess of electrons; the substance is called *n*-type (for "negative"). Another layer rests on top and is made of a material that has a dearth of electrons—or, as electrical engineers like to say, an excess of positively charged particles known as holes. This material is *p*-type (for "positive"). At the junction of the *n* and *p* layers is the so-called active layer, where light is emitted.

Applying a voltage drives electrons and holes into the active layer, where they meet. As they join, they emit photons—the basic units of light. The atomic structures of the active layer and adjoining materials on each side determine the number of photons produced and their wavelengths.

In early LEDs, made in the 1960s with a combination of gallium, arsenic and phosphorus to yield red light, electrons merged with holes relatively inefficiently: for every 1,000 electrons, only one red photon was produced. Such an LED generated less than one-tenth the amount of light found in a comparably powered, red-filtered incandescent bulb.

Over time, however, dramatic improvements in output were realized, especially at the red end of the spectrum. In 1999 Michael Krames and his coworkers at Hewlett-Packard set an efficiency record, building LEDs that transform more than 55 percent of the incoming electrons into photons at the red wavelength. Chief among the reasons for these improvements has been the continued rise in material quality and the development of substances that allow the efficient transformation of electrons and holes into photons. One of the biggest boosts in efficiency came when scientists found that the materials do not have to be homogeneous. Instead, each layer can have a different chemical makeup, so that when placed next to the active layer they can confine the electrons and holes better, thereby increasing the odds that an electron can combine with a hole to produce light.

Researchers have also learned to tailor the properties of the semiconducting layers. They can vary the material composition of the active layer and "dope" the layers with impurities. Doping alters the n- and p-type characteristics of the semiconductor layers. To take the simplest example, consider silicon. Silicon belongs to group IV of the periodic table: it has four electrons that are available for bonding with other atoms. In the crystal form, each atom shares electrons with neighboring atoms.

When a small number of group III atoms—that is, those containing three electrons in their outermost energy level, such as boron—are incorporated into the crystal, the resulting structure has an insufficient number of bonds to share with the surrounding silicon atoms. Vacancies for electrons appear, creating holes and rendering the material p-type.

Conversely, elements that belong to group V of the periodic table, such as phosphorus, have an extra electron in their outermost energy level. When silicon is doped with phosphorus, the crystal gains electrons, making the material n-type.

In LEDs, the crystal is not silicon but a mixture of group III and group V elements. By carefully controlling the concentration of aluminum, gallium, indium and phosphorus, for example, and by incorporating suitable dopants, typically tellurium and magnesium, researchers can control the formation of the n-side and the p-side, making LEDs that emit at the red, orange or yellow wavelengths. By the early 1970s red LEDs containing gallium arsenide phosphide were bright enough to illuminate the first calculators and digital clocks.

Another key to LED improvement lies in manufacturing techniques that reliably create viable, smooth crystals instead of lumpy, defect-riddled systems. The atomic lattices of the p and n materials must match up with those of the underlying supporting substrate and active layer. One such manufacturing method is vapor phase deposition, in which hot gases are channeled over a substrate to create a thin film. This technique was first incorporated into a high-volume LED manufacturing process at Monsanto in the late 1960s. In 1977 a different process of vapor phase deposition, one utilizing cool gases directed over a hot substrate, was demonstrated by Russell D. Dupuis, now at the University of Texas, to produce semiconductor lasers. This process, which enables the growth of a wider variety of materials, is now used to make high-quality LEDs. Shuji Nakamura, now at the University of California at Santa Barbara, used a variation of the technique to manufacture high-quality gallium nitride crystal capable of shining blue light. (For a profile of Nakamura, see "Blue Chip," by Glenn Zorpette; *Scientific American*, August 2000.)

In the mid-1990s a team at Hewlett-Packard found another way to enhance brightness—by reshaping the chip itself. Through careful manipulation, researchers can remove the original gallium arsenide wafer on which the active layer was grown, replace it with a transparent gallium phosphide wafer and sculpt an LED into the shape of an inverted pyramid. This shape decreases the number of internal reflections and thus boosts the amount of light escaping from the chip.

The Great White Hope

Thanks to these improvements in color and brightness, researchers have begun to zero in on making affordable, bright-white LEDs. Low-power white LEDs with an efficiency somewhat better than incandescent bulbs are already available commercially, but high-power devices suitable for illumination are still far too expensive to be mass-marketed. The potential benefits of such a

LEDs and Lasers

Although light-emitting diodes and laser diodes sound similar—in fact, both are made of semiconductor materials—they are very different beasts, designed to behave in different ways and to tackle different jobs.

Laser diodes take the form of a semiconductor material between what is essentially a pair of mirrors. The region between the mirrors is called the resonator cavity. When electricity goes through the semiconductor, it gives off photons, which then bounce back and forth inside the cavity, exciting other, nearby electron-hole pairs to release more photons at the same wavelength. The light increases continuously in intensity, with the photons marching in lockstep together as they oscillate between the two mirrors. If one of the mirrors allows just a small fraction of the light to escape, then some of the photons exit. All at the same wavelength and in phase, they produce an extremely narrow column of pure, bright light at a single wavelength. In physicists' terms, the photons are coherent.

This extremely well-defined beam is one of the main characteristics of a laser. As such, it is something like a scalpel: sharp, thin and able, with proper optics, to do delicate work, such as reading the fine pits on a compact disc or scanning the bar codes in a checkout line.

By comparison, the widely scattered light of an LED is like the patter of raindrops. Because LEDs are not in a proper cavity (that is, not between mirrors), the photons they emit are, in a sense, incoherent. Light comes out not in a unidirectional column but in a broader, more diffuse pattern, composed of a spread of wavelengths from one area of the spectrum. The photons an LED produces may not be all at the exact same wavelength, but they are close enough so that they are perceived by the human eye as being the same color.

LEDs Light the Deep

When marine biologist Greg Marshall of National Geographic Television wanted to film deep-diving animals such as sperm whales, he faced several problems. These creatures can plunge thousands of feet below the surface, to where it is virtually pitch-black and the pressures are enormous. Further compounding the situation is that any kind of visible lighting would affect his subjects' behavior, attracting or repelling them and their prey. By causing the whales to act abnormally, the standard underwater light would defeat the entire purpose of his project.

The solution: compact LEDs that emit light at the near-infrared wavelength, making for a light that the videotape can "see" but the animal cannot. When placed inside a hardy, torpedo-shaped metal cylinder containing an automatic camera, the devices act as invisible headlights, illuminating objects two or three meters away without altering the whales' behavior.

The small size of the LEDs meant that there was room to cram other equipment inside, such as devices to record audio, time of day, depth and duration of dive, direction, temperature and velocity. Dubbed Crittercam, the whole automatic camera package is small enough to be placed on the backs of whales, seals, dolphins, penguins, sea lions and other marine creatures, giving scientists a whale's-eye view of the sea. After filming, a time-release device in the harness lets go of the Crittercam, allowing it to bob to the surface where it can be retrieved, along with its precious footage.

Crittercam is the result of 14 years of experimentation, much of which was on suitable light sources. When the first blue LEDs came out commercially about four years ago—at $40 apiece—Marshall was one of the first to own one, which he purchased directly from the factory in Japan. (Unfortunately, he found that whereas blue-wavelength LEDs penetrated the deep ocean's gloom most efficiently, their light was too visible to his swimming subjects. Near-infrared became the choice instead. In other words, wavelength selection matters, a big advantage when it comes to LEDs.)

Marshall's effort paid off. Watching the footage from a Crittercam mounted to the dorsal fin of a shark is as exciting as watching the chase scene in a spy movie. And the camera has enabled marine biologists to observe behaviors never seen before, such as seals blowing bubbles and "singing" undersea as they perform courtship rituals.

—THE EDITORS

light, if made cheaply, are enormous. Instead of dealing with fragile, hot, gas-filled glass bulbs that burn out relatively quickly and waste most of their energy in the form of heat, consumers would own long-lasting, solid-state interior lights. In automo-biles, for example, the LEDs would last the life-time of the car. And the minimal power demands of LEDs mean that more energy is left in the auto-mobile's battery for all the onboard electronic devices.

Society as a whole could benefit as well. Lighting represents 20 to 30 percent of the U.S. electrical use, and even the best standard illumination systems convert no more than about 25 percent of electricity into light. If white LEDs could be made to match the efficiency of today's red LEDs, they could reduce energy needs and cut the amount of carbon dioxide pumped into the air by electrical generating plants by 300 megatons a year.

The first company to mass-produce affordable high-brightness white LEDs stands to capture an estimated $12 billion worldwide market for illumination lighting. That's why the big three players of lighting—Philips, Osram Sylvania and General Electric—are spending so much on LED research and development and why newer companies are springing up, such as LumiLeds, a joint venture between Philips and Agilent Technologies, for which one of us—Craford—is chief technology officer.

Low-power white LEDs are already used for cell-phone backlights and pedestrian walk signals. Second-generation, higher-power LEDs suitable for, say, landscape and accent lighting are becoming available. But large-scale replacement of lamps for general-purpose illumination are not expected for a decade or two because of the difficulty in making white LEDs efficient and cost-competitive.

There are two main ways to generate white light from LEDs. One way is to combine the output of LEDs at the red, green and blue wavelengths, based on the additive principle of color theory. The problem with this technique is that it is difficult to mix the colors of the LEDs efficiently with good uniformity and control.

The second way relies on an LED photon to excite a phosphor. For example, one can package a yellow phosphor around a blue LED. When the energy of the LED strikes the phosphor, it becomes excited and gives off yellow light, which mixes with the blue light from the LED to give white light. Alternatively, one can use an ultraviolet LED to excite a mixture of red, green and blue phosphors to give white light. This process, similar to that in fluorescent tubes, is simpler than mixing three colors but is inherently less efficient, because energy is lost in converting ultraviolet or blue light into lower-energy light (that is, light toward the red end of the spectrum). Moreover, light is also lost because of scattering and absorption in the phosphor packaging.

In any case, the high cost of LED chips and packages currently makes both approaches prohibitive for illumination applications. The best commercial white LEDs now cost about 50 cents per lumen, compared with a fraction of a penny per lumen for a typical incandescent bulb.

Whichever method is chosen—and both will most likely be important for different applications—the key issues are to reduce production costs substantially and to improve performance. Still, it may be a while before consumers accept LEDs, which cost more up front but are cheaper over the span of a decade. As energy prices rise and the consequences of global warming become more urgent, LEDs should become more attractive. One solution to our energy and environmental problems, it seems, may soon come to light.

—FEBRUARY 2001

Power-Thrifty PCs

Billion-dollar savings with better power supplies.

STEVEN ASHLEY

Putting a personal computer to sleep is typically the only means for users to conserve electricity, besides frequent, often inconvenient, shutdowns. Now a new focus of energy savings for the PC has emerged—its power supply.

When a PC is operating, its power supply typically converts only 60 to 70 percent of the 120-volt AC power into the 12-, 5- and 3.3-volt DC juice the internal system components need. The rest is mostly lost to heat. Each of the estimated 205 million PCs in the U.S. consumes an average of about 300 kilowatt-hours of power annually, and that figure does not include the monitor's energy usage. Making PC power supplies 80 percent efficient, researchers say, could shave U.S. energy use by 1 to 2 percent and pare $1 billion or more from the nation's yearly electric bills while cutting emissions from generating plants significantly.

That is the goal of new energy-saving efforts being undertaken by federal and state agencies, environmental groups, electric utilities and the computer industry. "In the past," says Craig W. Hershberg, a product development manager in the U.S. Environmental Protection Agency's Energy Star program, "we promoted greater use of instantly available 'sleep modes' to save PC energy use, but we've found that approach to be less than totally satisfactory, because it relies on the users to implement," many of whom do not bother to do so. Moreover, often home computer and entertain-

ment systems are networked and must stay on to be fully functional, which makes sleep-mode management difficult. Instead, Hershberg continues, "we're aiming at making the PC power supply more efficient—a target that doesn't require the user to do anything special."

Today's PCs use switching-mode power supplies (SMPS), says Michael Archer, chief technology officer at EOS, a division of Celetronix USA in Simi Valley, California. SMPS rely on a fast-acting switch to chop up the current, which is ultimately converted into low-voltage DC signals. Standard, "forced commutation" SMPS rely on a process "in which the current is made to turn on and off when it doesn't want to," Archer explains; in contrast, higher-efficiency "resonance-based" SMPS "only control the movement of that energy and so produce fewer losses." They can better match the demand for power with the supply and so produce less wasted energy.

In recent benchmark tests, the supplies that were 80 percent efficient cut energy use 15 to 25 percent across the board, reports Chris Calwell, director of policy and research for Ecos Consulting, a Portland, Oregon–based firm that promotes energy-efficient products. "Such improved units would cost about $5 more apiece wholesale but over four years of use would save about $25 in electricity costs." Ecos has formed partnerships with utilities to offer financial incentives to PC makers that install efficient power supplies.

Energy shavers are also targeting power-hungry central processing units (CPUs) and graphics cards. Intel and other chipmakers, for example, are now selling CPUs designed for laptops to desktop PC manufacturers. Laptop CPUs, designed to maximize battery life, can slow their processing speeds, thereby drawing less voltage. And engineers are looking for ways to improve the efficiency of the newest video cards, which may draw 50 to 60 watts each—as much as an entire computer.

Ecos and environmental watchdog Natural Resources Defense Council, working with Intel and others, have joined with the California Energy Commission and the EPA to launch a global competition to identify innovative design concepts that could boost efficiency (see www.efficientpower supplies.org). Researchers at Ecos meanwhile are developing performance metrics by which PCs can be assessed in the same way that miles per gallon measures automotive fuel usage—with a benchmark score divided by the system's electrical consumption. This metric could serve as the basis for new PC energy efficiency ratings.

—JUNE 2004

RENEWABLE ENERGY AND OTHER OPTIONS

Turning Green

Shell International projects a renewable energy future.

TIM BEARDSLEY

Prediction is always dangerous, and predicting the fortunes of energy sources is the riskiest form of this professional sport. Still, after many years in the field Shell has a better track record than most. The giant corporation's planning group is credited, for example, with alerting the company's management to the possibility of an oil crisis before the oil price hike of 1973. So when Shell talks (particularly when it talks to itself), everyone tries to listen. At the moment, knowledgeable ears are trained in the direction of the Shell International Petroleum Company in London, a service company for the Shell group.

And what they are hearing is definitely not orthodox stuff. Shell's business environment group, headed by Roger Rainbow, has sketched a future in which renewable sources will grow to dominate

world energy production by the year 2050. That perspective contrasts sharply with conservative studies by the World Energy Council (WEC), an international energy industry organization, and the International Energy Agency (IEA), an intergovernmental body. The WEC, for example, considers that "new" renewable sources, which include solar, wind, small hydroelectric, modern biomass and ocean sources, may account for only 5 percent of the world's energy output in 2020. In this view, fossil fuels will provide most of global energy needs through the middle of the next century, while nuclear fission plays an important supporting role.

The WEC's projections, the result of a three-year, $5 million study, were published last year. According to one of its midrange projections, annual global energy production will increase by

80 percent by 2020, to the equivalent of 16 billion metric tons of oil. That output will be needed to meet the needs of a human population that will be on its way from 5.5 billion (the 1990 figure) to 8.1 billion in 2020. (The numbers come from estimates by the United Nations and the World Bank.) The WEC and IEA studies presume that the new technologies will simply not have matured enough to capture a large fraction of the markets for coal, oil and natural gas.

But Rainbow and his colleagues, notably Georges DuPont-Roc, head of the planning group's energy division, disagree. Although the Shell exercise is not yet complete, Rainbow and DuPont-Roc have given officials at the World Bank and the U.S. Department of Energy a peek at the work in progress. And Peter Kassler of the Shell group has described some of the project's key aspects to the World Petroleum Council.

Kassler describes two possible geopolitical scenarios for the next 25 years. In one, the global trend toward economic liberalization and democratic reform in the 1980s continues to roll forward. That leads to a large increase in energy demand in developing countries, especially China and India, the world's most populous nations. At the same time, however, energy efficiency improves because of increased competition. Energy taxes internalize environmental costs, which help to stimulate the development of cleaner technologies.

Renewables gain importance in the second scenario, too, but less so, and in a distinctly grim setting. Regional economic and political tensions dominate the globe. Demand for oil increases, albeit slowly, but there is far less improvement in energy efficiency than in the first scheme. Protectionist policy and law weaken market forces. Oil price shocks exacerbate deteriorating international relations, and environmental anxieties spur government control of energy industries to ever-stricter levels. New markets for renewables are "largely in poor countries" or are developed locally and cheaply.

Kassler points out that the protectionist option is bad news not only for oil companies but also for the environment. Under any plausible view, developing countries account for most of the growth in demand over the next 30 years. If they do not gain access to new, energy-efficient technologies, they will follow the energy-inefficient path taken by the developed countries.

Under the more optimistic view, Kassler speculates, renewable energy technologies may well "start to be competitive with fossil fuels around 2020 or 2030." He points out that fossil fuel technologies will probably be unable to lower costs as quickly as will the younger upstarts. Then "potential uses of the new technologies would grow." Developing countries might leapfrog over the industrialized nations toward an energy-efficient future.

Kassler foresees changes on the demand side, too. Virtual reality might, he conjectures, lead to a reduction in the demand for travel, thus breaking the long-standing exponential growth in personal mobility. In any event, fossil fuel use would start to decline around the middle of next century.

Emissions of carbon dioxide, which most atmospheric scientists expect to lead to significant global warming, would start to fall. In the midrange futures that the WEC considers, carbon dioxide concentrations will continue to rise from the current level of about 358 parts per million throughout next century, reaching about 600 parts per million in 2100, while still rising.

Rainbow says he is convinced that, for the long term, business-as-usual scenarios are "fundamentally and deeply flawed." He believes it is "obvious" that "there won't be that much coal, oil and gas being used in 100 years." Shell has considered "green" energy futures in the past, but they assumed stringent environmental regulation. The new work is remarkable because it envisions a sustainable future without draconian controls, says Christopher Flavin, an energy analyst at the Worldwatch Institute in Washington, D.C.

Not everyone is convinced that Shell has got the future right. Lee Schipper, an energy researcher at Lawrence Berkeley Laboratory, notes that the company deliberately considers wide-ranging possibilities. Schipper, who was himself formerly in Shell's planning group, indicates that the new analysis is "not yet good enough to go on the record." Rainbow says he expects to air more details about his group's thinking later this year.

Even in its embryonic state the DuPont-Roc/Rainbow vision is applauded by environmentalists such as Flavin. He suggests that Shell's conclusions "discredit" the conservative World Energy Council predictions. Flavin believes that hydrogen generated by electrolyzing water using power from photovoltaic plants will be the fuel of the second half of the next century.

Nobody will predict the unfolding reality closely. But when a major oil company effectively projects the end of the fossil fuel age, it is a sure harbinger that we are moving into the future on fast-forward.

—SEPTEMBER 1994

The Rise of Renewable Energy

Solar cells, wind turbines and biofuels are poised to become major energy sources. New policies could dramatically accelerate that evolution.

DANIEL M. KAMMEN

No plan to substantially reduce greenhouse gas emissions can succeed through increases in energy efficiency alone. Because economic growth continues to boost the demand for energy—more coal for powering new factories, more oil for fueling new cars, more natural gas for heating new homes—carbon emissions will keep climbing despite the introduction of energy efficient vehicles, buildings and appliances. To counter the alarming trend of global warming, the U.S. and other countries must make a major commitment to developing renewable energy sources that generate little or no carbon.

Renewable energy technologies were suddenly and briefly fashionable three decades ago in response to the oil embargoes of the 1970s, but the interest and support were not sustained. In recent years, however, dramatic improvements in the performance and affordability of solar cells, wind turbines and biofuels—ethanol and other fuels derived from plants—have paved the way for mass commercialization. In addition to their

Overview

- Thanks to advances in technology, renewable sources could soon become large contributors to global energy.
- To hasten the transition, the U.S. must significantly boost its R&D spending on energy.
- The U.S. should also levy a fee on carbon to reward clean energy sources over those that harm the environment.

environmental benefits, renewable sources promise to enhance America's energy security by reducing the country's reliance on fossil fuels from other nations. What is more, high and wildly fluctuating prices for oil and natural gas have made renewable alternatives more appealing.

We are now in an era where the opportunities for renewable energy are unprecedented, making this the ideal time to advance clean power for decades to come. But the endeavor will require a long-term investment of scientific, economic and political resources. Policymakers and ordinary citizens must demand action and challenge one another to hasten the transition.

Let the Sun Shine

Solar cells, also known as photovoltaics, use semiconductor materials to convert sunlight into electric current. They now provide just a tiny slice of the world's electricity: their global generating capacity of 5,000 megawatts (MW) is only 0.15 percent of the total generating capacity from all sources. Yet sunlight could potentially supply 5,000 times as much energy as the world currently consumes. And thanks to technology improvements, cost declines and favorable policies in many states and nations, the annual production of pho-

tovoltaics has increased by more than 25 percent a year for the past decade and by a remarkable 45 percent in 2005. The cells manufactured last year added 1,727 MW to worldwide generating capacity, with 833 MW made in Japan, 353 MW in Germany and 153 MW in the U.S.

Solar cells can now be made from a range of materials, from the traditional multicrystalline silicon wafers that still dominate the market to thin-film silicon cells and devices composed of plastic or organic semiconductors. Thin-film photovoltaics are cheaper to produce than crystalline silicon cells but are also less efficient at turning light into power. In laboratory tests, crystalline cells have achieved efficiencies of 30 percent or more; current commercial cells of this type range from 15 to 20 percent. Both laboratory and commercial efficiencies for all kinds of solar cells have risen steadily in recent years, indicating that an expansion of research efforts would further enhance the performance of solar cells on the market.

Solar photovoltaics are particularly easy to use because they can be installed in so many places— on the roofs or walls of homes and office buildings, in vast arrays in the desert, even sewn into clothing to power portable electronic devices. The state of California has joined Japan and Germany in leading a global push for solar installations; the "Million Solar Roof" commitment is intended to create 3,000 MW of new generating capacity in the state by 2018. Studies done by my research group, the Renewable and Appropriate Energy Laboratory at the University of California, Berkeley, show that annual production of solar photovoltaics in the U.S. alone could grow to 10,000 MW in just 20 years if current trends continue.

The biggest challenge will be lowering the price of the photovoltaics, which are now relatively expensive to manufacture. Electricity produced by crystalline cells has a total cost of 20 to 25 cents per kilowatt-hour, compared with four to six cents for coal-fired electricity, five to seven cents for power produced by burning natural gas, and six to nine cents for biomass power plants. (The cost of nuclear power is harder to pin down because experts disagree on which expenses to include in the analysis; the estimated range is two to 12 cents per kilowatt-hour.) Fortunately, the prices of solar cells have fallen consistently over the past decade, largely because of improvements in manufacturing processes. In Japan, where 290 MW of solar generating capacity were added in 2005 and an even larger amount was exported, the cost of photovoltaics has declined 8 percent a year; in California, where 50 MW of solar power were installed in 2005, costs have dropped 5 percent annually.

Surprisingly, Kenya is the global leader in the number of solar power systems installed per capita (but not the number of watts added). More than 30,000 very small solar panels, each producing only 12 to 30 watts, are sold in that country annually. For an investment of as little as $100 for the panel and wiring, the system can be used to charge a car battery, which can then provide enough power to run a fluorescent lamp or a small black-and-white television for a few hours a day. More Kenyans adopt solar power every year than make connections to the country's electric grid. The panels typically use solar cells made of amorphous silicon; although these photovoltaics are only half as efficient as crystalline cells, their cost is so much lower (by a factor of at least four) that they are more affordable and useful for the two billion people worldwide who currently have no access to electricity. Sales of small solar power systems are booming in other African nations as well, and advances in low-cost photovoltaic manufacturing could accelerate this trend.

Furthermore, photovoltaics are not the only fast-growing form of solar power. Solar-thermal systems, which collect sunlight to generate heat, are also undergoing a resurgence. These systems have long

been used to provide hot water for homes or factories, but they can also produce electricity without the need for expensive solar cells. In one design, for example, mirrors focus light on a Stirling engine, a high-efficiency device containing a working fluid that circulates between hot and cold chambers. The fluid expands as the sunlight heats it, pushing a piston that, in turn, drives a turbine.

In the fall of 2005 a Phoenix company called Stirling Energy Systems announced that it was planning to build two large solar-thermal power plants in southern California. The company signed a 20-year power purchase agreement with Southern California Edison, which will buy the electricity from a 500-MW solar plant to be constructed in the Mojave Desert. Stretching across 4,500 acres, the facility will include 20,000 curved dish mirrors, each concentrating light on a Stirling engine about the size of an oil barrel. The plant is expected to begin operating in 2009 and could later be expanded to 850 MW. Stirling Energy Systems also signed a 20-year contract with San Diego Gas & Electric to build a 300-MW, 12,000-dish plant in the Imperial Valley. This facility could eventually be upgraded to 900 MW.

The financial details of the two California projects have not been made public, but electricity produced by present solar-thermal technologies costs between five and 13 cents per kilowatt-hour, with dish-mirror systems at the upper end of that range. Because the projects involve highly reliable technologies and mass production, however, the generation expenses are expected to ultimately drop closer to four to six cents per kilowatt-hour—that is, competitive with the current price of coal-fired power.

Blowing in the Wind

Wind power has been growing at a pace rivaling that of the solar industry. The worldwide generating capacity of wind turbines has increased more than 25 percent a year, on average, for the past decade, reaching nearly 60,000 MW in 2005. The growth has been nothing short of explosive in Europe—between 1994 and 2005, the installed wind power capacity in European Union nations jumped from 1,700 to 40,000 MW. Germany alone has more than 18,000 MW of capacity thanks to an aggressive construction program. The northern German state of Schleswig-Holstein currently meets one-quarter of its annual electricity demand with more than 2,400 wind turbines, and in certain months wind power provides more than half the state's electricity. In addition, Spain has 10,000 MW of wind capacity, Denmark has 3,000 MW, and Great Britain, the Netherlands, Italy and Portugal each have more than 1,000 MW.

In the U.S. the wind power industry has accelerated dramatically in the past five years, with total generating capacity leaping 36 percent to 9,100 MW in 2005. Although wind turbines now produce only 0.5 percent of the nation's electricity, the potential for expansion is enormous, especially in the windy Great Plains states. (North Dakota, for example, has greater wind energy resources than Germany, but only 98 MW of generating capacity is installed there.) If the U.S. constructed enough wind farms to fully tap these resources, the turbines could generate as much as 11 trillion kilowatt-hours of electricity, or nearly three times the total amount produced from all energy sources in the nation last year. The wind industry has developed increasingly large and efficient turbines, each capable of yielding four to six MW. And in many locations, wind power is the cheapest form of new electricity, with costs ranging from four to seven cents per kilowatt-hour.

The growth of new wind farms in the U.S. has been spurred by a production tax credit that provides a modest subsidy equivalent to 1.9 cents per kilowatt-hour, enabling wind turbines to compete with coal-fired plants. Unfortunately, Congress has repeatedly threatened to eliminate the tax credit. Instead of instituting a long-term subsidy for wind power, the lawmakers have extended the tax credit on a year-to-year basis, and the continual uncertainty has slowed investment in wind farms. Congress is also threatening to derail a proposed 130-turbine farm off the coast of Massachusetts that would provide 468 MW of generating capacity, enough to power most of Cape Cod, Martha's Vineyard and Nantucket.

The reservations about wind power come partly from utility companies that are reluctant to embrace the new technology and partly from so-called NIMBY-ism. ("NIMBY" is an acronym for Not in My Backyard.) Although local concerns over how wind turbines will affect landscape views may have some merit, they must be balanced

SEE *Figure 12 in color section.*

against the social costs of the alternatives. Because society's energy needs are growing relentlessly, rejecting wind farms often means requiring the construction or expansion of fossil fuel–burning power plants that will have far more devastating environmental effects.

Green Fuels

Researchers are also pressing ahead with the development of biofuels that could replace at least a portion of the oil currently consumed by motor vehicles. The most common biofuel by far in the U.S. is ethanol, which is typically made from corn and blended with gasoline. The manufacturers of ethanol benefit from a substantial tax credit: with the help of the $2 billion annual subsidy, they sold more than 16 billion liters of ethanol in 2005 (almost 3 percent of all automobile fuel by volume), and production is expected to rise 50 percent by 2007. Some policymakers have questioned the wisdom of the subsidy, pointing to studies showing

Plugging Hybrids

The environmental benefits of renewable biofuels would be even greater if they were used to fuel plug-in hybrid electric vehicles (PHEVs). Like more conventional gasoline-electric hybrids, these cars and trucks combine internal-combustion engines with electric motors to maximize fuel efficiency, but PHEVs have larger batteries that can be recharged by plugging them into an electrical outlet. These vehicles can run on electricity alone for relatively short trips; on longer trips, the combustion engine kicks in when the batteries no longer have sufficient juice. The combination can drastically reduce gasoline consumption: whereas conventional sedans today have a fuel economy of about 30 miles per gallon (mpg) and nonplug-in hybrids such as the Toyota Prius average about 50 mpg, PHEVs could get an equivalent of 80 to 160 mpg. Oil use drops still further if the combustion engines in PHEVs run on biofuel blends such as E85, which is a mixture of 15 percent gasoline and 85 percent ethanol.

If the entire U.S. vehicle fleet were replaced overnight with PHEVs, the nation's oil consumption would decrease by 70 percent or more, completely eliminating the need for petroleum imports. The switch would have equally profound implications for protecting the earth's fragile climate, not to mention the elimination of smog. Because most of the energy for cars would come from the electric grid instead of from fuel tanks, the environmental impacts would be concentrated in a few thousand power plants instead of in hundreds of millions of vehicles. This shift would focus the challenge of climate protection squarely on the task of reducing the greenhouse gas emissions from electricity generation.

PHEVs could also be the salvation of the ailing American auto industry. Instead of continuing to lose market share to foreign companies, U.S. automakers could become competitive again by retooling their factories to produce PHEVs that are significantly more fuel-efficient than the nonplug-in hybrids now sold by Japanese companies. Utilities would also benefit from the transition because most owners of PHEVs would recharge their cars at night, when power is cheapest, thus helping to smooth the sharp peaks and valleys in demand for electricity. In California, for example, the replacement of 20 million conventional cars with PHEVs would increase nighttime electricity demand to nearly the same level as daytime demand, making far better use of the grid and the many power plants that remain idle at night. In addition, electric vehicles not in use during the day could supply electricity to local distribution networks at times when the grid was under strain. The potential benefits to the electricity industry are so compelling that utilities may wish to encourage PHEV sales by offering lower electricity rates for recharging vehicle batteries.

Most important, PHEVs are not exotic vehicles of the distant future. DaimlerChrysler has already introduced a PHEV prototype, a plug-in hybrid version of the Mercedes-Benz Sprinter Van that has 40 percent lower gasoline consumption than the conventionally powered model. And PHEVs promise to become even more efficient as new technologies improve the energy density of batteries, allowing the vehicles to travel farther on electricity alone.

that it takes more energy to harvest the corn and refine the ethanol than the fuel can deliver to combustion engines. In a recent analysis, though, my colleagues and I discovered that some of these studies did not properly account for the energy content of the by-products manufactured along with the ethanol. When all the inputs and outputs were correctly factored in, we found that ethanol has a positive net energy of almost five megajoules per liter.

We also found, however, that ethanol's impact on greenhouse gas emissions is more ambiguous. Our best estimates indicate that substituting corn-based ethanol for gasoline reduces greenhouse gas emissions by 18 percent, but the analysis is hampered by large uncertainties regarding certain agricultural practices, particularly the environmental costs of fertilizers. If we use different assumptions about these practices, the results of switching to ethanol range from a 36 percent drop in emissions to a 29 percent increase. Although corn-based ethanol may help the U.S. reduce its reliance on foreign oil, it will probably not do much to slow global warming unless the production of the biofuel becomes cleaner.

But the calculations change substantially when the ethanol is made from cellulosic sources: woody plants such as switchgrass or poplar. Whereas most makers of corn-based ethanol burn fossil fuels to provide the heat for fermentation, the producers of cellulosic ethanol burn lignin—an unfermentable part of the organic material—to heat the plant sugars. Burning lignin does not add any greenhouse gases to the atmosphere, because the emissions are offset by the carbon dioxide absorbed during the growth of the plants used to make the ethanol. As a result, substituting cellulosic ethanol for gasoline can slash greenhouse gas emissions by 90 percent or more.

Another promising biofuel is so-called green diesel. Researchers have produced this fuel by first gasifying biomass—heating organic materials enough

that they release hydrogen and carbon monoxide—and then converting these compounds into long-chain hydrocarbons using the Fischer-Tropsch process. (During World War II, German engineers employed these chemical reactions to make synthetic motor fuels out of coal.) The result would be an economically competitive liquid fuel for motor vehicles that would add virtually no greenhouse gases to the atmosphere. Oil giant Royal Dutch/Shell is currently investigating the technology.

The Need for R&D

Each of these renewable sources is now at or near a tipping point, the crucial stage when investment and innovation, as well as market access, could enable these attractive but generally marginal providers to become major contributors to regional and global energy supplies. At the same time, aggressive policies designed to open markets for renewables are taking hold at city, state and federal levels around the world. Governments have adopted these policies for a wide variety of reasons: to promote market diversity or energy security, to bolster industries and jobs, and to protect the environment on both the local and global scales. In the U.S. more than 20 states have adopted standards setting a minimum for the fraction of electricity that must be supplied with renewable sources. Germany plans to generate 20 percent of its electricity from renewables by 2020, and Sweden intends to give up fossil fuels entirely.

Even President George W. Bush said, in his now famous State of the Union address this past January, that the U.S. is "addicted to oil." And although Bush did not make the link to global warming, nearly all scientists agree that humanity's addiction to fossil fuels is disrupting the earth's climate. The time for action is now, and at last the tools exist to alter energy production and consumption in ways that simultaneously benefit the economy and the environment. Over the past 25 years, however, the public and private funding of research and development in the energy sector has withered. Between 1980 and 2005 the fraction of all U.S. R&D spending devoted to energy declined from 10 to 2 percent. Annual public R&D funding for energy sank from $8 billion to $3 billion (in 2002 dollars); private R&D plummeted from $4 billion to $1 billion [see sidebar, "R&D Is Key," below].

To put these declines in perspective, consider that in the early 1980s energy companies were investing more in R&D than were drug companies, whereas today investment by energy firms is an order of magnitude lower. Total private R&D funding for the entire energy sector is less than that of a single large biotech company. (Amgen, for example, had R&D expenses of $2.3 billion in 2005.) And as R&D spending dwindles, so does

innovation. For instance, as R&D funding for photovoltaics and wind power has slipped over the past quarter of a century, the number of successful patent applications in these fields has fallen accordingly. The lack of attention to long-term research and planning has significantly weakened our nation's ability to respond to the challenges of climate change and disruptions in energy supplies.

Calls for major new commitments to energy R&D have become common. A 1997 study by the President's Committee of Advisors on Science and Technology and a 2004 report by the bipartisan National Commission on Energy Policy both recommended that the federal government double its R&D spending on energy. But would such an expansion be enough? Probably not. Based on assessments of the cost to stabilize the amount of carbon dioxide in the atmosphere and other

U.S. R&D Spending in the Energy Sector

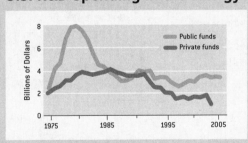

R&D Is Key

Spending on research and development in the U.S. energy sector has fallen steadily since its peak in 1980. Studies of patent activity suggest that the drop in funding has slowed the development of renewable energy technologies. For example, the number of successful patent applications in photovoltaics and wind power has plummeted as R&D spending in these fields has declined.

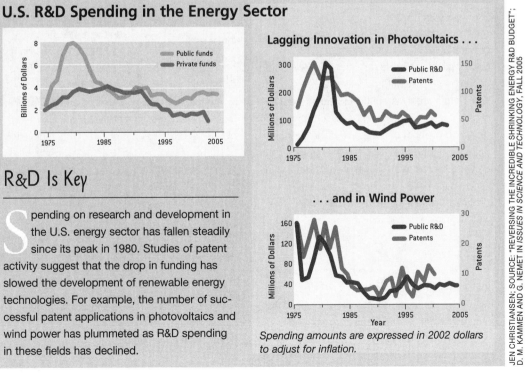

Spending amounts are expressed in 2002 dollars to adjust for inflation.

JEN CHRISTIANSEN; SOURCE: "REVERSING THE INCREDIBLE SHRINKING ENERGY R&D BUDGET"; D. M. KAMMEN AND G. NEMET IN *ISSUES IN SCIENCE AND TECHNOLOGY,* FALL 2005

The Least Bad Fossil Fuel

Although renewable energy sources offer the best way to radically cut greenhouse gas emissions, generating electricity from natural gas instead of coal can significantly reduce the amount of carbon added to the atmosphere. Conventional coal-fired power plants emit 0.25 kilogram of carbon for every kilowatt-hour generated. (More advanced coal-fired plants produce about 20 percent less carbon.) But natural gas (CH_4) has a higher proportion of hydrogen and a lower proportion of carbon than coal does. A combined-cycle power plant that burns natural gas emits only about 0.1 kilogram of carbon per kilowatt-hour (see *graph*).

Unfortunately, dramatic increases in natural gas use in the U.S. and other countries have driven up the cost of the fuel. For the past decade, natural gas has been the fastest-growing source of fossil fuel energy, and it now supplies almost 20 percent of America's electricity. At the same time, the price of natural gas has risen from an average of $2.50 to $3 per million Btu in 1997 to more than $7 per million Btu today.

The price increases have been so alarming that in 2003, then Federal Reserve Board Chair Alan Greenspan warned that the U.S. faced a natural gas crisis. The primary solution proposed by the White House and some in Congress was to increase gas production. The 2005 Energy Policy Act included large subsidies to support gas producers, increase exploration and expand imports of liquefied natural gas (LNG). These measures, however, may not enhance energy security, because most of the imported LNG would come from some of the same OPEC countries that supply petroleum to the U.S. Furthermore, generating electricity from even the cleanest natural gas power plants would still emit too much carbon to achieve the goal of keeping carbon dioxide in the atmosphere below 450 to 550 parts per million by volume. (Higher levels could have disastrous consequences for the global climate.)

Improving energy efficiency and developing renewable sources can be faster, cheaper and cleaner and provide more security than developing new gas supplies. Electricity from a wind farm costs less than that produced by a natural gas power plant if the comparison factors in the full cost of plant construction and forecasted gas prices. Also, wind farms and solar arrays can be built more rapidly than large-scale natural gas plants. Most critically, diversity of supply is America's greatest ally in maintaining a competitive and innovative energy sector. Promoting renewable sources makes sense strictly on economic grounds, even before the environmental benefits are considered.

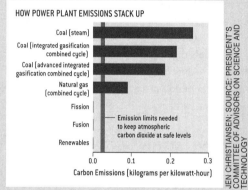

HOW POWER PLANT EMISSIONS STACK UP

Coal (steam)
Coal (integrated gasification combined cycle)
Coal (advanced integrated gasification combined cycle)
Natural gas (combined cycle)
Fission
Fusion
Renewables

— Emission limits needed to keep atmospheric carbon dioxide at safe levels

0.0   0.1   0.2   0.3
Carbon Emissions (kilograms per kilowatt-hour)

JEN CHRISTIANSEN; SOURCE: PRESIDENT'S COMMITTEE OF ADVISORS ON SCIENCE AND TECHNOLOGY

studies that estimate the success of energy R&D programs and the resulting savings from the technologies that would emerge, my research group has calculated that public funding of $15 billion to $30 billion a year would be required—a fivefold to 10-fold increase over current levels.

Greg F. Nemet, a doctoral student in my laboratory, and I found that an increase of this magnitude would be roughly comparable to those that occurred during previous federal R&D initiatives such as the Manhattan Project and the Apollo program, each of which produced demonstrable economic benefits in addition to meeting its objectives. American energy companies could also boost their R&D spending by a factor of 10, and it would still be below the average for U.S. industry overall. Although government funding is essential to supporting early-stage technologies, private-sector R&D is the key to winnowing the best ideas and reducing the barriers to commercialization.

Raising R&D spending, though, is not the only way to make clean energy a national priority. Educators at all grade levels, from kindergarten to college, can stimulate public interest and activism by teaching how energy use and production affect the social and natural environment. Nonprofit organizations can establish a series of contests that would reward the first company or private group to achieve a challenging and worthwhile energy goal, such as constructing a building or appliance that can generate its own power or developing a commercial vehicle that can go 200 miles on a single gallon of fuel. The contests could be modeled after the Ashoka awards for pioneers in public policy and the Ansari X Prize for the developers of space vehicles. Scientists and entrepreneurs should also focus on finding clean, affordable ways to meet the energy needs of people in the developing world. My colleagues and I, for instance, recently detailed the environmental benefits of improving cooking stoves in Africa.

But perhaps the most important step toward creating a sustainable energy economy is to institute market-based schemes to make the prices of carbon fuels reflect their social cost. The use of coal, oil and natural gas imposes a huge collective toll on society, in the form of health-care expenditures for ailments caused by air pollution, military spending to secure oil supplies, environmental damage from mining operations, and the potentially devastating economic impacts of global warming. A fee on carbon emissions would provide a simple, logical and transparent method to reward renewable, clean energy sources over those that harm the economy and the environment. The tax revenues could pay for some of the social costs of carbon emissions, and a portion could be designated to compensate low-income families who spend a larger share of their income on energy. Furthermore, the carbon fee could be combined with a cap-and-trade program that would set limits on carbon emissions but also allow the cleanest energy suppliers to sell permits to their dirtier competitors. The federal government has used such programs with great success to curb other pollutants, and several northeastern states are already experimenting with greenhouse gas–emissions trading.

Best of all, these steps would give energy companies an enormous financial incentive to advance the development and commercialization of renewable energy sources. In essence, the U.S. has the opportunity to foster an entirely new industry. The threat of climate change can be a rallying cry for a clean-technology revolution that would strengthen the country's manufacturing base, create thousands of jobs and alleviate our international trade deficits—instead of importing foreign oil, we can export high-efficiency vehicles, appliances, wind turbines and photovoltaics. This transformation can turn the nation's energy sector into something that was once deemed impossible: a vibrant, environmentally sustainable engine of growth.

—SEPTEMBER 2006

Solar Energy

Technology will allow radiation from the sun to provide nonpolluting and cheap fuels, as well as electricity.

WILLIAM HOAGLAND

Every year the earth's surface receives about 10 times as much energy from sunlight as is contained in all the known reserves of coal, oil, natural gas and uranium combined. This energy equals 15,000 times the world's annual consumption by humans. People have been burning wood and other forms of biomass for thousands of years, and that is one way of tapping solar energy. But the sun also provides hydropower, wind power and fossil fuels—in fact, all forms of energy other than nuclear, geothermal and tidal.

Attempts to collect the direct energy of the sun are not new. In 1861 a mathematics instructor named Augustin-Bernard Mouchot of the Lycée de Tours in France obtained the first patent for a solar-powered motor. Other pioneers also investigated using the sun's energy, but the convenience of coal and oil was overwhelming. As a result, solar power was mostly forgotten until the energy crisis of the 1970s threatened many major economies.

Economic growth depends on energy use. By 2025 the worldwide demand for fuel is projected to increase by 30 percent and that for electricity by 265 percent. Even with more efficient use and conservation, new sources of energy will be required. Solar energy could provide 60 percent of the electricity and as much as 40 percent of the fuel.

Extensive use of more sophisticated solar energy technology will have a beneficial impact on air pollution and global climatic change. In developing countries, it can alleviate the environmental damage caused by the often-inefficient practice of burning plant material for cooking and heating. Advanced solar technologies have the potential to use less land than does biomass cultivation: photosynthesis typically captures less than 1 percent of the available sunlight, but modern solar technologies can, at least in the laboratory, achieve efficiencies of 20 to 30 percent. With such efficiencies, the U.S. could meet its current demand for energy by devoting less than 2 percent of its land area to energy collection.

It is unlikely that a single solar technology will predominate. Regional variations in economics and the availability of sunlight will naturally favor some approaches over others. Electricity may be generated by burning biomass, erecting wind turbines, building solar-powered heat engines, laying out photovoltaic cells or harnessing the energy in rivers with dams. Hydrogen fuel can be produced by electrochemical cells or biological processes—involving microorganisms or enzymes—that are driven by sunlight. Fuels such as ethanol and methanol may be generated from biomass or other solar technologies.

Solar energy also exists in the oceans as waves and gradients of temperature and salinity, and they, too, are potential reservoirs to tap. Unfortunately, although the energy stored is enormous, it is diffuse and expensive to extract.

Growing Energy

Agricultural or industrial wastes such as wood chips can be burned to generate steam for turbines. Such facilities are competitive with conventional electricity production wherever biomass is cheap. Many such plants already exist, and more are being commissioned. Recently in Värnamo, Sweden, a modern power plant using gasified wood to fuel a jet engine was completed. The facility converts 80 percent of the energy in the wood to provide six megawatts of power and nine megawatts of heat for the town. Although biomass combustion can be polluting, such technology makes it extremely clean.

Progress in combustion engineering and biotechnology has also made it economical to convert plant material into liquid or gaseous fuels. Forest products, "energy crops," agricultural residues and other wastes can be gasified and used to synthesize methanol. Ethanol is released when sugars, derived from sugarcane or various kinds of grain crops or from wood (by converting cellulose), are fermented.

Alcohols are now being blended with gasoline to enhance the efficiency of combustion in car engines and to reduce harmful tailpipe emissions. But ethanol can be an effective fuel in its own right, as researchers in Brazil have demonstrated. It may be cost-competitive with gasoline by 2000. In the future, biomass plantations could allow such energy to be "grown" on degraded land in developing nations. Energy crops could also allow for better land management and higher profits. But much research is needed to achieve consistently high crop yields in diverse climates.

Questions do remain as to how useful biomass can be, even with technological innovations.

Photosynthesis is inherently inefficient and requires large supplies of water. A 1992 study commissioned by the United Nations concluded that 55 percent of the world's energy needs could be met by biomass by 2050. But the reality will hinge on what other options are available.

Wind Power

Roughly 0.25 percent of the sun's energy reaching the lower atmosphere is transformed into wind—a minuscule part of the total but still a significant source of energy. By one estimate, 80 percent of the electrical consumption in the U.S. could be met by the wind energy of North and South Dakota alone. The early problems surrounding the reliability of "wind farms" have now been by and large resolved, and in certain locations the electricity produced is already cost-competitive with conventional generation.

In areas of strong wind—an average of more than 7.5 meters per second—electricity from wind farms costs as little as $0.04 per kilowatt-hour. The cost should drop to below $0.03 per kilowatt-hour by the year 2000. In California and Denmark more than 17,000 wind turbines have been completely integrated into the utility grid. Wind now supplies about 1 percent of California's electricity.

One reason for the reduction is that stronger and lighter materials for the blades have allowed wind machines to become substantially larger. The turbines now provide as much as 0.5 megawatt apiece. Advances in variable-speed turbines have reduced stress and fatigue in the moving parts, thus improving reliability. Over the next 20 years better materials for air foils and transmissions and smoother controls and electronics for handling high levels of electrical power should become available.

One early use of wind energy will most likely be for islands or other areas that are far from an electrical grid. Many such communities currently import diesel for generating power, and some are actively seeking alternatives. By the middle of the next century, wind power could meet 10 to 20 percent of the world's demand for electrical energy.

The major limitation of wind energy is that it is intermittent. If wind power constitutes more than 25 to 45 percent of the total power supply, any shortfall causes severe economic penalties. Better means of energy storage would allow the percentage of wind power used in the grid to increase substantially. (I will return to this question presently.)

Heat Engines

One way of generating electricity is to drive an engine with the sun's radiant heat and light. Such solar-thermal electric devices have four basic components; namely, a system for collecting sunlight, a receiver for absorbing it, a thermal storage device and a converter for changing the heat to electricity. The collectors come in three basic configurations: a parabolic dish that focuses light to a point, a parabolic trough that focuses light to a line and an array of flat mirrors spread over several acres that reflect light onto a single central tower.

These devices convert between 10 and 30 percent of the direct sunlight to electricity. But uncertainties remain regarding their life span and reliability. A particular technical challenge is to develop a Stirling engine that performs well at low cost. (A Stirling engine is one in which heat is added continuously from the outside to a gas contained in a closed system.)

Solar ponds, another solar-thermal source, contain highly saline water near their bottom. Typically, hot water rises to the surface, where it cools off. But salinity makes the water dense, so that hot water can stay at the bottom and thus retain its heat. The pond traps the sun's radiant heat, creating a high temperature gradient. Hot, salty fluid is drawn out from the bottom of the pond and allowed to evaporate; the vapor is used to drive a Rankine-cycle engine similar to that

installed in cars. The cool liquid at the top of the pond can also be used, for air-conditioning.

A by-product of this process is freshwater from the steam. Solar ponds are limited by the large amounts of water they need and are more suited to remote communities that require freshwater as well as energy. Use of solar ponds has been widely investigated in countries with hot, dry climates, such as in Israel.

Solar Cells

The conversion of light directly to electricity by the photovoltaic effect was first observed by the French physicist Edmond Becquerel in 1839. When photons shine on a photovoltaic device, commonly made of silicon, they eject electrons from their stable positions, allowing them to move freely through the material. A voltage can then be generated using a semiconductor junction. A method of producing extremely pure crystalline silicon for photovoltaic cells with high voltages and efficiencies was developed in the 1940s. It proved to be a tremendous boost for the industry. In 1958 photovoltaics were first used by the American space program to power the radio of the *U.S. Vanguard I* space satellite with less than one watt of electricity.

Although significant advances have been made in the past 20 years—the current record for photovoltaic efficiency is more than 30 percent—cost remains a barrier to widespread use. There are two approaches to reducing the high price: producing cheap materials for so-called flat-plate systems, and using lenses or reflectors to concentrate sunlight onto smaller areas of (expensive) solar cells. Concentrating systems must track the sun and do not use the diffuse light caused by cloud cover as efficiently as flat-plate systems. They do, however, capture more light early and late in the day.

Virtually all photovoltaic devices operating today are flat-plate systems. Some rotate to track

A New Chance for Solar Energy

Solar power is getting cheaper—in fact, the cost of filching the sun's rays has fallen more than 65 percent in the past 10 years. It has not become inexpensive enough, though, to rival fossil fuels, so solar energy remains a promising, not yet fully mature alternative. Sales run only about $1 billion annually, as opposed to roughly $800 billion for standard sources, and solar customers still generally reside in isolated areas, far from power grids.

But a new proposal from an American utility may well make solar power conventional—or at least more competitive. Enron Corporation, the largest U.S. supplier of natural gas, recently joined forces with Amoco Corporation, owner of the photovoltaic cell producer Solarex. The two companies intend to build a 100-megawatt solar plant in the Nevada desert by the end of 1996. The facility, which could supply a city of 100,000, will initially sell energy for 5.5 cents a kilowatt-hour—about three cents cheaper on average than the electricity generated by oil, coal or gas. "If they can pull this off, it can revolutionize the whole industry," comments Robert H. Williams of Princeton University. "If they fail, it is going to set back the technology 10 years."

Despite its magnitude, the $150 million plan does not mean that the solar age has finally dawned: Enron's low price is predicated on tax exemptions from the Department of Energy and on guaranteed purchases by the federal government. Nor does it mark a sudden technological breakthrough: Solarex manufactures a conventional thin-film, silicon-based photovoltaic cell that is able to transform into electricity about 8 percent of the sunlight that reaches it. Rather the significance of Enron's venture—should the bid be accepted by the government—is that it paves the way for other companies to make large-scale investments in solar power.

Such investments could bring the price of solar-power technology and delivery down even further—for both large, grid-based markets and for the more dispersed, off-the-grid markets that are the norm in many developing countries. "This marks a shift in approach," explains Nicholas Lenssen, formerly at the Worldwatch Institute in Washington, D.C., and now at E Source in Boulder, Colorado. "It allows them to attract lower-risk, long-term capital, not just venture capital, which is very costly." Which all means the Nevada desert may soon be home to a very different, but still very hot, kind of test site.

—THE EDITORS

the sun, but most have no moving parts. One may be optimistic about the future of these devices because commercially available efficiencies are well below theoretical limits and because modern manufacturing techniques are only now being applied.

Photovoltaic electricity produced by either means should soon cost less than $0.10 per kilowatt-hour, becoming competitive with conventional generation early in the next century.

Storing Energy

Sunlight, wind and hydropower all vary intermittently, seasonally and even daily. Demand for energy fluctuates as well; matching supply and demand can be accomplished only with storage. A study by the Department of Energy estimated that by 2030 in the U.S., the availability of appropriate storage could enhance the contribution of renewable energy by about 18 quadrillion British thermal units per year.

With the exception of biomass, the more promising long-term solar systems are designed to produce only electricity. Electricity is the energy carrier of choice for most stationary applications, such as heating, cooling, lighting and machinery. But it is not easily stored in suitable quantities. For use in transportation, lightweight, high-capacity energy storage is needed. Sunlight can also be used to produce hydrogen fuel. The technologies required to do so directly (without generating electricity first) are in the very early stages of development but in the long term may prove the best.

Sunlight falling on an electrode can produce an electric current to split water into hydrogen and oxygen, by a process called photoelectrolysis. The term "photobiology" is used to describe a whole class of biological systems that produce hydrogen. Even longer-term research may lead to photocatalysts that allow sunlight to split water directly into its component substances.

When the resulting hydrogen is burned as a fuel or is used to produce electricity in a fuel cell, the only by-product is water. Apart from being environmentally benign, hydrogen provides a way to alleviate the problem of storing solar energy. It can be held efficiently for as long as required. Over distances of more than 1,000 kilometers, it costs less to transport hydrogen than to transmit electricity. Residents of the Aleutian Islands have developed plans to make electricity from wind turbines, converting it to hydrogen for storage. In addition, improvements in fuel cells have allowed a number of highly efficient, nonpolluting uses of hydrogen to be developed, such as electric vehicles powered by hydrogen.

A radical shift in our energy economy will require alterations in the infrastructure. When the decision to change is made will depend on the importance placed on the environment, energy security or other considerations. In the U.S., federal programs for research into renewable energy have been on a rollercoaster ride. Even the fate of the Department of Energy is uncertain.

At present, developed nations consume at least 10 times the energy per person than is used in developing countries. But the demand for energy is rising fast everywhere. Solar technologies could enable the developing world to skip a generation of infrastructure and move directly to a source of energy that does not contribute to global warming or otherwise degrade the environment. Developed countries could also benefit by exporting these technologies—if additional incentives are at all necessary for investing in the future of energy from the sun.

—SEPTEMBER 1995

More Power to Solar

Solar cells were one of Scientific American's *top fifty technology trends for 2005. Photovoltaic advances make the ever-lagging technology more of a competitor.*

GEORGE MUSSER

Brazilians joke that theirs is the country of the future—and always will be. Likewise, solar power has always been the ultimate green technology of the future. But maybe the sun is finally rising. The photovoltaic market, though small, has been growing briskly: by more than 60 percent in 2004. Plastering your roof with solar cells now runs as little as 20 cents per kilowatt-hour over the system's estimated lifetime, which is approaching what most households pay for electricity.

One especially promising technology that emerged in the 1990s was to make solar cells from plastic spiked with nanometer-scale crystals. Even those composite devices, though, were restricted to absorbing visible light. This year a group led by Edward H. Sargent at the University of Toronto coaxed them to absorb infrared light as well. A concoction of lead sulfide particles a few nanometers in size can absorb wavelengths as long as two microns. Thus able to harvest a wider swath of the solar spectrum, inexpensive plastic cells could rival the performance of pricey silicon ones.

Other avant-garde photovoltaic devices consist of nanoparticles coated with dye and doused in electrolyte, an approach pioneered by Michael Grätzel of the Swiss Federal Institute of Technology in Lausanne a decade ago. The dye handles the job of absorbing photons and generating a current of electrons. Because the source of the electrons (the dye) is divorced from the matrix through which they flow in (the electrolyte) and out (the nanoparticles), electrons are less likely to get prematurely recaptured by atoms, a process that impairs current flow in conventional cells. Consequently, the dye-based cells work better under weak lighting conditions.

Tsutomu Miyasaka and Takurou N. Murakami of the Toin University of Yokohama have extended the technique to create the world's first photocapacitor: a solar cell that both generates and stores electricity. Alongside the dye-coated particles, the researchers slapped down layers of activated carbon, which traps electrons and holds them until a switch completes the circuit. Under a 500-watt bulb, their latest design takes a couple of minutes to charge up to 0.8 volt. It has a capacitance of about 0.5 farad per square centimeter, which would give a typical solar panel the same energy storage capacity as the so-called ultracapacitors developed to replace or supplement batteries in hybrid cars and uninterruptible power supplies. In 2004 Miyasaka founded a company, Peccell Technologies, to commercialize this and other innovations.

Another way to store energy is in the form of hydrogen gas. In the late 1960s Japanese researchers Akira Fujishima and Kenichi Honda discovered that a solar cell can act like an artificial tree leaf, splitting water into its constituent elements. The trouble was that the materials involved, such as titanium dioxide, absorb mostly ultraviolet light. Restricted to such a narrow band of spectrum, the process was pitifully inefficient. Tinkering with their chemical properties allowed the cells to absorb visible light but also made them prone to corrosion.

Grätzel recently developed a way around this unhappy trade-off: put two solar cells together. The first contains tungsten trioxide or iron oxide, which soaks up the ultraviolet. The second is one of his dye-sensitized cells, which absorbs the rest of the visible spectrum and provides more electrons to aid the photolysis.

A year ago Hydrogen Solar, a British company trying to commercialize the work, made the announcement of a nearly 10-fold improvement in the efficiency of water splitting. It estimates that hydrogen produced this way would still cost about twice as much as hydrogen from natural gas but might become competitive if greenhouse gas emissions were restricted. You wouldn't need to go to a gas station to refill your fuel-cell car; the solar panel on the roof of your house could be your private gas station.

—DECEMBER 2005

Blowing Out to Sea

Offshore wind farms may finally reach the U.S.

WENDY WILLIAMS

With little alteration to the national power grid, the U.S. could quickly get at least 12 percent of its electricity from wind. Yet currently, wind generators supply only about 0.5 percent, in part because people don't want to live underneath the tall turbines. In Europe one solution to the people problem is to place the wind machines out at sea, where the winds are stronger anyway.

Acknowledging this potential, a Yarmouth, Massachusetts, company plans to build America's first offshore wind farm by the end of 2005. Cape Wind Associates has slated construction of a 420-megawatt wind project on a shallow sandbar known as Horseshoe Shoal, located five miles south of Cape Cod between the islands of Nantucket and Martha's Vineyard. It would be the world's second largest, after Ireland's recently proposed 520-megawatt farm.

Each of the 170 ultra-high-tech wind turbines will stand 260 feet tall at the turbine hub, and each blade will be up to 150 feet long. The turbines, which should be visible in the distance from the Hyannisport Kennedy enclave, will be laid out in a grid pattern over 25 square miles of saltwater. An underwater cable will run from the turbine complex to a Cape Cod substation. Project developers claim that at peak operation the farm will satisfy almost all the electricity needs of Cape residents—a critical selling point in a region that suffers increasingly from air inversions and smog.

Less than a decade old, offshore wind technology has been virtually ignored by U.S. companies until now. In Europe, though, it's the hot idea in "green" energy. Denmark, for example, trumpets the fact that 50 percent of its energy will come from wind by 2030. If successful, offshore wind farms could solve many problems encountered with land-based wind technology in densely populated regions. Ocean winds are stronger and steadier. Land acquisition is unnecessary. And perhaps most important, the huge turbines are out of sight and earshot of most people. Initially, fishermen worried about their catch volume decreasing, but several European studies suggest that the heavily anchored turbines act like shipwrecks and in fact improve fish numbers.

On the flip side, investment costs are mammoth. Cape Wind, having already invested several million dollars in planning studies, expects to spend a total of $600 million. James S. Gordon, president of Cape Wind, is confident that the whole package can be financed through private sources. Under his 27-year leadership, Energy Management, a partner in Cape Wind, has built a number of natural gas–fired plants in New England. Says Gordon: "We're creating a national model for America's energy and environmental future."

The U.S. Department of Energy is "watching the Cape project very closely," remarks Brian Parsons, a scientist with the DOE's National Renewable Energy Laboratory. But the size of the undertaking has raised some eyebrows. "I'd be a little skeptical about starting with something that big," warns wind-farm engineering expert Tim Cockerill, a research fellow at the University of Sunderland in England. Others in Europe, however, are thinking along the same lines as Cape Wind. Researchers at the Dutch Offshore Wind Energy Converter project are aiming for a single six-megawatt offshore turbine by 2008. Continued interest may prove within the decade whether this alternative to fossil fuels is more than just a passing gust.

—MARCH 2002

When Blade Meets Bat

Unexpected bat kills threaten future wind farms.

WENDY WILLIAMS

The interaction of bats and wind turbines is emerging as a major and unexpected problem in northern Appalachia. From mid-August through October 2003, during the fall migration period, at least 400 bats died at FPL Energy's 44-turbine Mountaineer Wind Energy Center on Backbone Mountain in West Virginia.

The bats apparently died by colliding with the wind turbines, but why so many animals were killed at this particular site remains a mystery. The public outcry over these numbers threatens to delay or halt construction of some of the additional several hundred wind turbines planned for the tristate region of West Virginia, western Maryland and south-central Pennsylvania.

Steve Stengel, a spokesperson for FPL, which is based in Juno Beach, Florida, says the company is cooperating with federal biologists to study the problem of bat kills at Mountaineer. "We don't know exactly why it happened," he states. "We're moving quickly to find out as much as we can." Some scientists believe that the migrating bats may not be using their echolocation when the collisions occur. Others speculate that the wind turbines may be emitting high-pitched sounds that draw the bats to the site. Still others suggest that the animals may be getting caught in wind shear associated with the turning turbines.

West Virginia biologists have identified the majority of the 400 bats that were recovered from the Mountaineer site: mostly common species such as red bats, eastern pipistrelles and hoary bats. "What's scary," remarks biologist Albert Manville of the U.S. Fish and Wildlife Service, "is that we may be finding only a small percentage of what's been killed." That is because bats are very small and difficult to find in the field; also, scavengers could discover the bat corpses before researchers do.

At issue is the length of time that wind-energy entrepreneurs are devoting to preconstruction wildlife studies. The Fish and Wildlife Service issued voluntary siting guidelines last summer, indicating that a census of wildlife activity should precede the building of a wind farm. Some biologists feel that such a census should last two years, although some energy companies believe this length of time to be excessive. (The guidelines are voluntary because in many cases the federal agency has little enforcement power unless an endangered or threatened animal is actually killed.)

Concerned that the endangered Indiana bat may be at risk at FPL's 20-turbine wind project in Meyersdale, Pennsylvania, wildlife advocates are threatening legal action. They allege that thorough habitat studies were not done in advance of construction at Meyersdale.

A letter last October from a bat biologist hired by the project's builders would appear to back them up. Pennsylvania State University's Michael R. Gannon spent two days last spring looking for bat caves on the future wind-farm site. He suggested that Indiana bats may use the site as a summer habitat and noted that at least a summer-long study might be appropriate. But industry biologists disagreed, Gannon says. "A two-year study should have been conducted prior to the installation of the turbines to determine the potential risk to bats," he wrote in his letter. "Unless and until these data are available, it should be assumed that this site is a light path of the Indiana bats and that Indiana bats will be killed. . . . Data that are available indicate this as a very likely scenario."

FPL, which bought the project during development, still wants more information. "We are reviewing the matter," Stengel comments, "and after our review we will respond, if appropriate." When green energy meets red bats, the mammals seem to lose. Some wind farms are finding this species of bat, as well as many others, dead on their properties. Such discoveries could threaten planned wind farms and force revisions in the way turbines are sited.

—FEBRUARY 2004

Plan B for Energy

If efficiency improvements and incremental advances in today's technologies fail to halt global warming, could revolutionary new carbon-free energy sources save the day? Don't count on it—but don't count it out, either.

W. WAYT GIBBS

To keep this world tolerable for life as we like it, humanity must complete a marathon of technological changes whose finish line is far over the horizon. Robert H. Socolow and Stephen W. Pacala of Princeton University have compared the feat to a multigenerational relay race [see "A Plan to Keep Carbon in Check," page 52]. They outline a strategy to win the first 50-year leg by reining back carbon dioxide emissions from a century of unbridled acceleration. Existing technologies, applied both wisely and promptly, should carry us to this first milestone without trampling the global economy. That is a sound plan A.

The plan is far from foolproof, however. It depends on societies ramping up an array of carbon-reducing practices to form seven "wedges," each of which keeps 25 billion tons of carbon in the ground and out of the air. Any slow starts or early plateaus will pull us off track. And some scientists worry that stabilizing greenhouse gas emissions will require up to 18 wedges by 2056, not the seven that Socolow and Pacala forecast in their most widely cited model [see sidebar, "Plan B: Sooner—or Later?" page 212].

It is a mistake to assume that carbon releases will rise more slowly than will economic output and

energy use, argues Martin I. Hoffert, a physicist at New York University. As oil and gas prices rise, he notes, the energy industry is "re-carbonizing" by turning back to coal. "About 850 coal-fired power plants are slated to be built by the U.S., China and India—none of which signed the Kyoto Protocol," Hoffert says. "By 2012 the emissions of those plants will overwhelm Kyoto reductions by a factor of five."

Even if plan A works and the teenagers of today complete the first leg of the relay by the time they retire, the race will be but half won. The baton will then pass in 2056 to a new generation for the next and possibly harder part of the marathon: cutting the rate of CO_2 emissions in half by 2106.

Sooner or later the world is thus going to need a plan B: one or more fundamentally new technologies that together can supply 10 to 30 terawatts without belching a single ton of carbon dioxide. Energy buffs have been kicking around many such wild ideas since the 1960s. It is time to get serious about them. "If we don't start now building the infrastructure for a revolutionary change in the energy system," Hoffert warns, "we'll never be able to do it in time."

But what to build? The survey that follows sizes up some of the most promising options, as

Plan B: Sooner—or Later?

Staving off catastrophic global warming means bridging a gap between the amount of carbon emitted by business as usual and a flat path toward a stable carbon dioxide concentration. That gap may grow much more rapidly than Robert H. Socolow of Princeton and many economists typically estimate, warns NYU physicist Martin I. Hoffert. The standard "seven-wedge" scenario [see sidebar, "15 Ways to Make a Wedge," page 56] assumes that both the energy consumed per dollar of GDP and the carbon emitted per kilowatt of energy will continue to fall. Hoffert points out, however, that China and India have begun "recarbonizing," emitting more CO_2 per kilowatt every year as they build coal-fired plants. Carbon-to-energy ratios have stopped falling in the U.S. as well. Socolow acknowledges that the seven-wedge projection assumes substantial advances in efficiency and renewable energy production as part of business as usual. Even if those assumptions all prove correct, revolutionary technologies will still be needed to knock down carbon emissions in the latter half of the 21st century.

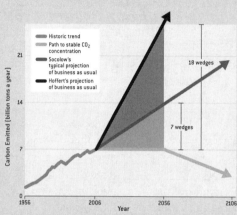

Historic trend
Path to stable CO_2 concentration
Socolow's typical projection of business as usual
Hoffert's projection of business as usual

18 wedges

7 wedges

Carbon Emitted [billion tons a year]

21

14

7

0
1956   2006   2056   2106
Year

JEN CHRISTIANSEN; SOURCES: ROBERT H. SOCOLOW AND MARTIN I. HOFFERT

well as a couple that are popular yet implausible. None of them is a sure thing. But from one of these ideas might emerge a new engine of human civilization.

Each technology is ranked by its estimated technical feasibility, ranging from 1 (implausible) to 5 (ready for market).

Nuclear Fusion

Reality Factor 3: Starry-eyed physicists point to the promise of unlimited fuel and minimal waste. But politicians blanch at fusion's price tag and worry about getting burned.

Fusion reactors—which make nuclear power by joining atoms rather than splitting them—top almost everyone's list of ultimate energy technologies for humanity. By harnessing the same strong thermonuclear force that fires the sun, a fusion plant could extract a gigawatt of electricity from just a few kilograms of fuel a day. Its hydrogen-isotope fuel would come from seawater and lithium, a common metal. The reactor would produce no greenhouse gases and relatively small amounts of low-level radioactive waste, which would become harmless within a century. "Even if the plant were flattened [by an accident or attack], the radiation level one kilometer outside the fence would be so small that evacuation would not be necessary," says Farrokh Najmabadi, a fusion expert who directs the Center for Energy Research at the University of California, San Diego.

The question is whether fusion can make a large contribution to the 21st century or is a 22nd-century solution. "A decade ago some scientists questioned whether fusion was possible, even in the lab," says David E. Baldwin, who as head of the energy group at General Atomics oversees the largest fusion reactor in the U.S., the DIII-D. But the past 20 years have seen dramatic improvements in tokamaks, machines that use giant electromagnetic coils to confine the ionized fuel within a doughnut-shaped chamber as it heats the plasma to more than 100 million degrees Celsius.

"We now know that fusion will work," Baldwin says. "The question is whether it is economically practical"—and, if so, how quickly fusion could move from its current experimental form into large-scale commercial reactors. "Even with a crash program," he says, "I think we would need 25 to 30 years" to develop such a design.

Fast Facts

FUSION REACTION:
NEXT-GENERATION FUSION REACTORS

Project	Place	Online
EAST	China	2006
SST-1	India	2006
K-Star	Korea	2008
NIF	U.S.	2009
ITER	France	2016
NCT	Japan	?

So far political leaders have chosen to push fusion along much more slowly. Nearly 20 years after it was first proposed, the International Thermonuclear Experimental Reactor (ITER) is only now nearing final approval. If construction begins on schedule next year, the $10 billion reactor should begin operation in southeastern France in 2016.

Meanwhile, an intermediate generation of tokamaks now nearing completion in India, China and Korea will test whether coils made of superconducting materials can swirl the burning plasma within its magnetic bottle for minutes at a time. Current reactors manage a few dozen seconds at best before their power supplies give out.

ITER aims for three principal goals. First, it must demonstrate that a large tokamak can control the fusion of the hydrogen isotopes deuterium and tritium into helium long enough to generate 10 times the energy it consumes. A secondary aim is to test ways to use the high-speed neutrons created by the reaction to breed tritium fuel—for example, by shooting them into a surrounding blanket of lithium. The third goal is to integrate the wide range of technologies needed for a commercial fusion plant.

If ITER succeeds, it will not add a single watt to the grid. But it will carry fusion past a milestone that nuclear fission energy reached in 1942, when Enrico Fermi oversaw the first self-sustaining nuclear chain reaction. Fission reactors were powering submarines 11 years later. Fusion is an incomparably harder problem, however, and some veterans in the field predict that 20 to 30 years of experiments with ITER will be needed to refine designs for a production plant.

Najmabadi is more optimistic. He leads a working group that has already produced three rough designs for commercial fusion reactors. The latest, called ARIES-AT, would have a more compact footprint—and thus a lower capital cost—

than ITER. The ARIES-AT machine would produce 1,000 megawatts at a price of roughly five cents per kilowatt-hour, competitive with today's oil- and gas-fired plants. If work on a commercial plant began in parallel with ITER, rather than decades after it goes online, fusion might be ready to scale up for production by midcentury, Najmabadi argues.

Fusion would be even more cost-competitive, Hoffert suggests, if the fast neutrons produced by tokamaks were used to transmute thorium (which is relatively abundant) into uranium (which may be scarce 50 years hence) to use as fuel in nuclear fission plants. "Fusion advocates don't want to sully its clean image," Hoffert observes, "but fusion-fission hybrids may be the way to go."

High-Altitude Wind

Reality Factor 4: The most energetic gales soar far over the tops of today's turbines. New designs would rise higher—perhaps even to the jet stream.

Wind is solar energy in motion. About 0.5 percent of the sunlight entering the atmosphere is transmuted into the kinetic energy of air: a mere 1.7 watts, on average, in the atmospheric column above every square meter of the earth. Fortunately, that energy is not distributed evenly but concentrated into strong currents. Unfortunately, the largest, most powerful and most consistent currents are all at high altitude. Hoffert estimates that roughly two-thirds of the total wind energy on this planet resides in the upper troposphere, beyond the reach of today's wind farms.

Ken Caldeira of the Carnegie Institution of Washington once calculated how wind power varies with altitude, latitude and season. The mother lode is the jet stream, about 10,000 meters (33,000 feet) up between 20 and 40 degrees latitude in the Northern Hemisphere. In the skies over the U.S., Europe, China and Japan—indeed, many of the countries best prepared to exploit it—wind power surges to 5,000 or even 10,000 watts a square meter. The jet stream does wander. But it never stops.

If wind is ever to contribute terawatts to the global energy budget, engineers will have to invent affordable ways to mine the mother lode. Three high-flying designs are in active development.

Magenn Power in Ottawa, Ontario, plans to begin selling next year a rotating, helium-filled generator that exploits the Magnus effect (best known for giving loft to spinning golf balls) to float on a tether up to 122 meters above the ground. The bus-size device will produce four kilowatts at its ground station and will retail for about $10,000—helium not included. The company aims to produce higher-flying, 1.6-megawatt units, each the size of a football field, by 2010.

"We looked at balloons; the drag they produce seemed unmanageable in high winds," says Al Grenier of Sky WindPower in Ramona, California. Grenier's venture is instead pursuing autogiros, which catch the wind with helicopter-like rotors. Rising to 10,000 meters, the machines could realize 90 percent of their peak capacity. The inconstancy of surface winds limits ground turbines to about half that. But the company has struggled to gather the $4 million it needs for a 250-kilowatt prototype.

Fast Facts

- Wind power capacity, currently about 58 gigawatts, is expected to triple by 2014.
- Helium-filled generators have to be refilled every few months.
- Number of tethered aerostats monitoring the U.S. border: eight.

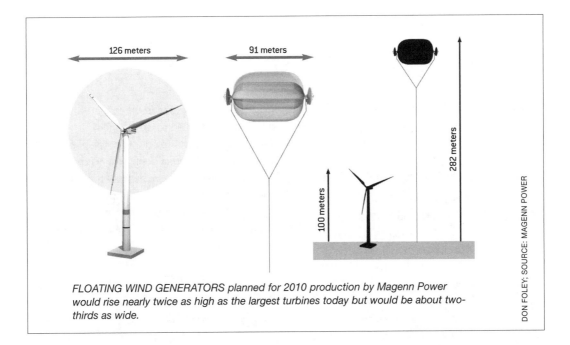

126 meters

91 meters

282 meters

100 meters

FLOATING WIND GENERATORS planned for 2010 production by Magenn Power would rise nearly twice as high as the largest turbines today but would be about two-thirds as wide.

DON FOLEY; SOURCE: MAGENN POWER

Still in the conceptual stages is the "laddermill," designed by astronaut Wubbo J. Ockels and his students at the Delft University of Technology in the Netherlands. Ockels envisions a series of computer-controlled kites connected by a long tether. The ladder of kites rises and descends, turning a generator on the ground as it yo-yos up and down. Simulations of the system suggest that a single laddermill reaching to the jet stream could produce up to 50 megawatts of energy.

Until high-altitude machines are fielded, no one can be certain how well they will hold up under turbulence, gusts and lightning strikes. Steep maintenance costs could be their downfall.

There are regulatory hurdles to clear as well. Airborne wind farms need less land than their terrestrial counterparts, but their operators must persuade national aviation agencies to restrict aircraft traffic in the vicinity. There is precedent for this, Grenier points out: the U.S. Air Force has for years flown up to a dozen large tethered aerostats at high altitude above the country's southern border.

By the standards of revolutionary technologies, however, high-altitude wind looks relatively straightforward and benign.

Space-Based Solar

Reality Factor 3: With panels in orbit, where the sun shines brightest—and all the time—solar could really take off. But there's a catch.

When Peter Glaser proposed in 1968 that city-size satellites could harvest solar power from deep space and beam it back to the earth as invisible microwaves, the idea seemed pretty far out, even given Glaser's credentials as president of the International Solar Energy Society. But after the oil crises of the 1970s sent fuel prices skyrocketing, NASA engineers gave the scheme a long hard look. The technology seemed feasible until, in 1979, they estimated the "cost to first power": $305

Sci-Fi Solutions

Reality Factor 1: Futuristic visions make for great entertainment. Too bad about the physics.

COLD FUSION AND BUBBLE FUSION

B. Stanley Pons and Martin Fleischmann spun a tempest in a teacup in 1989 with their claim of room-temperature fusion in a bottle. The idea drew a coterie of die-hard supporters, but mainstream scientists have roundly rejected that variety of cold fusion.

Theoretically more plausible—but still experimentally contentious—is sonofusion. In 2002 Rusi Taleyarkhan, a physicist then at Oak Ridge National Laboratory, reported in *Science* that beaming high-intensity ultrasound and neutrons into a vat of acetone caused microscopic bubbles to form and then implode at hypersonic speeds. The acetone had been made using deuterium, a neutron-bearing form of hydrogen, and Taleyarkhan's group claimed that the extraordinary temperatures and pressures created inside the imploding bubbles forced a few deuterium atoms to fuse with incoming neutrons to form tritium (hydrogen with two neutrons per atom). Another group at Oak Ridge replicated the experiment but saw no clear signs of fusion.

Taleyarkhan moved to Purdue University and continued reporting success with sonic fusion even as others tried but failed. Purdue this year investigated allegations that Taleyarkhan had interfered with colleagues whose work seemed to contradict his own. The results of the inquiry were sealed—and with them another chapter in the disappointing history of cold fusion. Other researchers hold out hope that different methods might someday turn a new page on sonofusion.

MATTER-ANTIMATTER REACTORS

The storied *Enterprise* starships fueled their warp drives with a mix of matter and antimatter; why can't we? The combination is undoubtedly powerful: a kilogram of each would, through their mutual annihilation, release about half as much energy as all the gasoline burned in the U.S. last year. But there are no known natural sources of antimatter, so we would have to synthesize it. And the most efficient antimatter maker in the world, the particle accelerator at CERN near Geneva, would have to run nonstop for 100 trillion years to make a kilogram of antiprotons.

So even though physicists have ways to capture the odd antiatom [see "Making Cold Antimatter," by Graham P. Collins; *Scientific American*, June 2005], antimatter power plants will never materialize.

billion (in 2000 dollars). That was the end of that project.

Solar and space technologies have made great strides since then, however, and space solar power (SSP) still has its champions. Hoffert cites two big advantages that high-flying arrays could lord over their earthbound brethren. In a geostationary orbit well clear of the earth's shadow and atmosphere, the average intensity of sunshine is eight times as strong as it is on the ground. And with the sun always in their sights, SSP stations could feed a reliable, fixed amount of electricity into the grid. (A rectifying antenna, or "rectenna," spread over several square kilometers of land could convert microwaves to electric current with about 90 percent efficiency, even when obstructed by clouds.)

"SSP offers a truly sustainable, global-scale and emission-free electricity source," Hoffert argues. "It is more cost-effective and more technologically fea-

Showstoppers

- Large teams of robots will have to work together to assemble the giant arrays.
- The microwave beams could cause interference with communications systems.
- Space agencies will have to boost their launch rates by a factor of about 80.
- Rectennas will occupy large swaths of land.

sible than controlled thermonuclear fusion." Yet there is minimal research funding for space-based solar, he complains, while a $10 billion fusion reactor has just been approved.

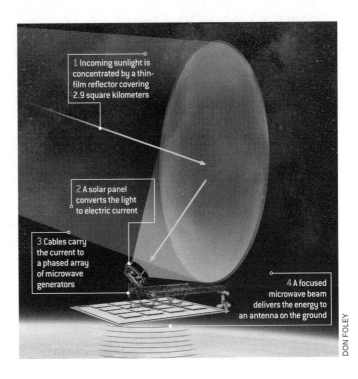

Giant solar collector in geosynchronous orbit would work day and night, in any weather. A pilot plant of the size to the right would intercept four gigawatts of sunlight, convert it to 1.8 GW of microwaves, and deliver 1.1 GW of electricity to the grid.

1 Incoming sunlight is concentrated by a thin-film reflector covering 2.9 square kilometers

2 A solar panel converts the light to electric current

3 Cables carry the current to a phased array of microwave generators

4 A focused microwave beam delivers the energy to an antenna on the ground

DON FOLEY

NASA did in fact fund small studies from 1995 to 2003 that evaluated a variety of SSP components and architectures. The designs took advantage of thin-film photovoltaics to create the electricity, high-temperature superconductors to carry it, and infrared lasers (in place of microwave emitters) to beam it to ground stations. Such high-tech innovations enabled SSP engineers to cut the systems' weight and thus reduce the formidable cost of launching them into orbit.

But here's the catch: the power-to-payload ratio, at a few hundred watts per kilogram, has remained far too low. Until it rises, space-based solar will never match the price of other renewable energy sources, even accounting for the energy storage systems that ground-based alternatives require to smooth over nighttime and poor-weather lulls.

Technical advances could change the game rapidly, however. Lighter or more efficient photovoltaic materials are in the works [see "Nanotech Solar Cells," this page]. In May, for example, researchers at the University of Neuchatel in Switzerland reported a new technique for depositing amorphous silicon cells on a space-hardy film that yields power densities of 3,200 watts per kilogram. Although that is encouraging, says John C. Mankins, who led NASA's SSP program from 1995 to 2003, "the devil is in the supporting structure and power management." Mankins sees more promise in advanced earth-to-orbit space transportation systems, now on drawing boards, that might cut launch costs from more than $10,000 a kilogram to a few hundred dollars in coming decades.

JAXA, the Japanese space agency, last year announced plans to launch by 2010 a satellite that will unfurl a large solar array and beam 100 kilowatts of microwave or laser power to a receiving station on the earth. The agency's long-term road map calls for flying a 250-megawatt prototype system by 2020 in preparation for a gigawatt-class commercial SSP plant a decade later.

NASA once had similarly grand designs, but the agency largely halted work on SSP when its priorities shifted to space exploration two years ago.

Nanotech Solar Cells

Reality Factor 4: Materials engineered from the atoms up could boost photovoltaic efficiencies from pathetic to profitable.

Five gigawatts—a paltry 0.038 percent of the world's consumption of energy from all sources. That, roughly, is the cumulative capacity of all photovoltaic (PV) power systems installed in the world, half a century after solar cells were first commercialized. In the category of greatest unfulfilled potential, solar-electric power is a technology without rival.

Even if orbiting arrays [see "Space-Based Solar," on page 215] never get off the ground, nanotechnology now looks set to rescue solar from its perennial irrelevance, however. Engineers are working on a wide range of materials that outshine the bulk silicon used in most PV cells today, improving both their efficiency and their cost.

The most sophisticated (and expensive) second-generation silicon cells eke out about 22 percent efficiency. New materials laced with quantum dots might double that, if discoveries reported this past March pan out as hoped. The dots, each less than 10 billionths of a meter wide, were created by groups at the National Renewable Energy Laboratory in Colorado and Los Alamos National Laboratory in New Mexico.

When sunlight hits a silicon cell, most of it ends up as heat. At best, a photon can knock loose one electron. Quantum dots can put a wider range of wavelengths to useful work and can kick out as many as seven electrons for every photon. Most of those electrons soon get stuck again, so engineers are testing better ways to funnel them into wires. They are also hunting for dot materials that are more environmentally friendly than the lead, selenium and cadmium in today's nanocrystals. Despite their high-tech name, the dots are relatively inexpensive to make.

Nanoparticles of a different kind promise to help solar compete on price. Near San Francisco, Nanosolar is building a factory that will churn out 200 million cells a year by printing nanoscopic bits of copper-indium-gallium-diselenide onto continuous reels of ultrathin film. The particles self-assemble into light-harvesting structures. Nanosolar's CEO says he is aiming to bring the cost down to 50 cents a watt.

The buzz has awakened energy giants. Shell now has a subsidiary making solar cells, and BP in June launched a five-year project with the California Institute of Technology. Its goal: high-efficiency solar cells made from silicon nanorods.

A Global Supergrid

Reality Factor 2: Revolutionary energy sources need a revolutionary superconducting electrical grid that spans the planet.

"A basic problem with renewable energy sources is matching supply and demand," Hoffert observes. Supplies of sunshine, wind, waves and even biofuel crops fade in and out unpredictably, and they tend to be concentrated where people are not. One solution is to build long-distance transmission lines from superconducting wires. When chilled to near absolute zero, these conduits can wheel tremendous currents over vast distances with almost no loss.

In July the BOC Group in New Jersey and its partners began installing 350 meters of superconducting cable into the grid in Albany, New York. The nitrogen-cooled link will carry up to 48 megawatts' worth of current at 34,500 volts. "We know the technology works; this project will demonstrate that," says Ed Garcia, a vice president at BOC.

At a 2004 workshop, experts sketched out designs for a "SuperGrid" that would simultane-

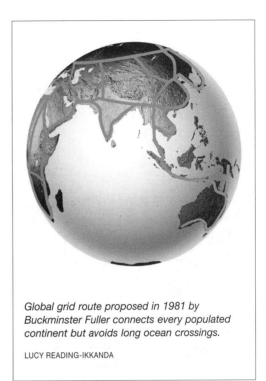

Global grid route proposed in 1981 by Buckminster Fuller connects every populated continent but avoids long ocean crossings.

LUCY READING-IKKANDA

ously transport electricity and hydrogen. The hydrogen, condensed to a liquid or ultracold gas, would cool the superconducting wires and could also power fuel cells and combustion engines [see "A Power Grid for the Hydrogen Economy," page 132].

With a transcontinental SuperGrid, solar arrays in Australia and wind farms in Siberia might power lights in the U.S. and air conditioners in Europe. But building such infrastructure would most likely take generations and trillions of dollars.

Waves and Tides

Reality Factor 5: The surging ocean offers a huge, but virtually untapped, energy resource. Companies are now gearing up to catch the wave.

The tide has clearly turned for the dream of harnessing the incessant motion of the sea. "Ocean energy is about 20 years behind wind power," acknowledges Roger Bedard, ocean energy leader at the Electric Power Research Institute. "But it certainly isn't going to take 20 years to catch up."

Through the 1980s and 1990s, advocates of tidal and wave power could point to only two commercial successes: a 240-megawatt (MW) tidal plant in France and a 20-MW tidal station in Nova Scotia. Now China has jumped onboard with a 40-kilowatt (kW) facility in Daishan. Six 36-kW turbines are soon to start spinning in New York City's East River. This summer the first commercial wave farm will go online in Portugal. And investors and governments are hatching much grander schemes.

The grandest is in Britain, where analysts suggest ocean power could eventually supply one-fifth of the country's electricity and fulfill its obligations under the Kyoto Protocol. The UK government in July ordered a feasibility study for a 16-kilometer dam across the Severn estuary, whose tides rank second largest in the world. The Severn barrage, as it is called, would cost $25 billion and produce 8.6 gigawatts when tides were flowing. Proponents claim it would operate for a century or more.

Environmental groups warn that the barrage would wreak havoc on the estuarine ecosystem. Better than a dam, argues Peter Fraenkel of Marine Current Turbines, would be arrays of the SeaGen turbines his company has developed. Such tide farms dotting the UK coast could generate almost as much electricity as the Severn dam but with less capital investment, power variation and environmental impact.

Fraenkel's claims will be put to a small test this year, when a tidal generator the company is installing in Strangford Lough begins contributing an average power of 540 kW to the grid in Northern Ireland. The machine works much like an underwater windmill, with two rotors sharing a single mast cemented into the seabed.

"The biggest advantage of tidal power is that it is completely predictable," Bedard says. "But on a global scale, it will never be very large." There are too few places where tides move fast enough.

Energetic waves are more capricious but also more ubiquitous. An analysis by Bedard's group found that if just 20 percent of the commercially viable offshore wave resources in the U.S. were harnessed with 50 percent–efficient wave farms, the energy produced would exceed all conventional hydroelectric generation in the country.

Four companies have recently completed sea trials of their wave conversion designs. One of them, Ocean Power Delivery, will soon begin reaping 2.25 MW off the coast of Portugal from three of its 120-meter-long Pelamis machines. If all goes well, it will order another 30 this year. Surf's up.

Designer Microbes

Reality Factor 4: Genetic engineers think they can create synthetic life-forms that will let us grow energy as easily as we do food.

"We view the genome as the software, or even the operating system, of the cell," said J. Craig Venter.

It's time for an upgrade, he suggested. Venter was preaching to the choir: a large group of biologists at the Synthetic Biology 2.0 conference this past May. Many of the scientists there have projects to genetically rewire organisms so extensively that the resulting cells would qualify as synthetic species. Venter, who gained fame and fortune for the high-speed methods he helped to develop to sequence the human genome, recently founded a company, Synthetic Genomics, to commercialize custom-built cells. "We think this field has tremendous potential to replace the petrochemical industry, possibly within a decade," he said.

That assessment may be overly optimistic; no one has yet assembled a single cell from scratch. But Venter reported rapid progress on his team's efforts to create artificial chromosomes that contain just the minimum set of genes required for self-sustaining life within a controlled, nutrient-rich environment. "The first synthetic prokaryotic cell [lacking a nucleus] will definitely happen within the next two years," he predicted. "And synthetic eukaryotic genomes [for cells with nuclei] will happen within a decade at most."

Venter envisions novel microbes that capture carbon dioxide from the smokestack of a power plant and turn it into natural gas for the boiler. "There are already thousands, perhaps millions, of organisms on our planet that know how to do this," Venter said. Although none of those species may be suited for life in a power plant, engineers could borrow their genetic circuits for new creations. "We also have biological systems under construction that are trying to produce hydrogen directly from sunlight, using photosynthesis," he added.

Steven Chu, director of Lawrence Berkeley National Laboratory, announced that his lab is readying a proposal for a major project to harness the power of the sun and turn it into fuels for transportation. With the tools of genetic engineering, Chu explained, "we can work on modifying plants and algaes to make them self-fertilizing and resistant to drought and pests." The novel crops would offer high yields of cellulose, which man-made microbes could then convert to fuels. Chu expects biological processing to be far more efficient than the energy-intensive processes, such as steam explosion and thermal hydrolysis, currently used to make ethanol.

With oil prices approaching $80 a barrel, bio-processing may not have to wait for life-forms built from scratch. GreenFuel in Cambridge, Massachusetts, has installed algae farms at power plants to convert up to 40 percent of the CO_2 they spew into raw material for biofuels. The company claims that a large algae farm next to a 1-GW plant could yield 50 million gallons a year of ethanol. "There are great opportunities here," Chu avers. "And not only that—it will help save the world."

SEPTEMBER 2006

ABOUT THE AUTHORS*

Roger N. Anderson is director of petroleum technology research at the Energy Center of Columbia University. After growing up with a father in the oil industry, Anderson completed his Ph.D. in earth sciences at the Scripps Institution of Oceanography at the University of California, San Diego. He sits on the board of directors of Bell Geospace and 4D Systems and spends his summers consulting for oil and service companies. Anderson has published more than 150 peer-reviewed scientific papers and holds seven U.S. patents.

Luis Miguel Ariza is a freelance science writer based in Madrid.

Steven Ashley is a *Scientific American* technology writer and editor.

Tim Beardsley is a *Scientific American* staff writer.

Ralph G. Bennett has played a leading role in the U.S. nuclear energy program. Currently director of nuclear energy at the U.S. Department of Energy's Idaho National Engineering and Environmental Laboratory (INEEL), he is a member of the team that leads the DOE's Generation IV effort.

Christopher E. Borroni-Bird joined GM in June 2000 as director of design and technology fusion, a group that applies emerging technology to improve vehicle design. He is also director of the AUTOnomy program, which includes the Hy-wire prototype vehicle. Together with J. Byron McCormick and Lawrence D. Burns, he plays a leading role in the fuel-cell development efforts of General Motors. Previously, Borroni-Bird managed Chrysler's Jeep Commander fuel-cell vehicle program.

Lawrence D. Burns is vice president of GM Research & Development and Planning. He oversees the company's advanced technology and innovation programs and is responsible for the company's product portfolio, capacity and business plans. Burns is a member of the Automotive Strategy Board, GM's highest-level management team. Together with J. Byron McCormick and Christopher E. Borroni-Bird he plays a leading role in the fuel-cell development efforts of General Motors.

Colin J. Campbell has worked in the oil industry for more than 40 years. After completing his Ph.D. in geology at the University of Oxford, he worked for Texaco as an exploration geologist and then at Amoco as chief geologist for Ecuador. His decade-long study of global oil-production trends has led to two books and numerous papers. He is currently associated with Petroconsultants in Geneva.

M. George Craford, who made the first yellow LED, managed the LED technology groups at Monsanto and Hewlett-Packard before becoming chief technology officer at LumiLeds Lighting in San Jose, California, a firm jointly created by Philips and Agilent Technologies to seek emerging LED applications.

John M. Deutch is an institute professor at the Massachusetts Institute of Technology and was a cochair of the 2003 interdisciplinary MIT study entitled *The Future of Nuclear Power*. He has held many government positions. He was director of energy research and undersecretary of energy (1977–1980) and later deputy secretary of defense (1994–1995) and director of central intelligence (1994–1996). He is currently cochairing an MIT study on the future of coal.

Rodger Doyle is a regular writer for *Scientific American*.

Baldur Eliasson, head of ABB's Energy and Global Change Program, is the Swiss representative to as well as vice chairman of the International Energy Agency's Greenhouse Gas Research and Development Program.

Safaa A. Fouda received a doctorate in chemical engineering from the University of Waterloo in 1976. Since 1981 she has worked at the CANMET Energy Technology Center, a Canadian government laboratory in Nepean, Ontario. There she manages a group of researchers studying natural gas conversion, emissions control, waste oil recycling and liquid fuels from renewable sources. Recently, she headed an international industrial consortium intent on developing better methods to convert natural gas to liquid fuels.

W. Wayt Gibbs is a former senior writer for *Scientific American*.

Paul M. Grant worked for IBM for 40 years, starting in 1953 at age 17 as a pinsetter at the company bowling alley. After earning a Ph.D. in physics at Harvard University, he joined the San Jose Research Laboratory, where he participated in the discovery of high-temperature superconductivity. From 1993 to 2004, Grant was a science fellow at the Electric Power Research Institute (EPRI), which was founded by coauthor Chauncey Starr in 1973.

William H. Hannum served as head of nuclear physics development and reactor safety research at the U.S. Department of Energy. He was also deputy director general of the Nuclear Energy Agency of the Organization for Economic Cooperation and Development in Paris.

James Hansen is director of the NASA Goddard Institute for Space Studies and a researcher at the Columbia University Earth Institute. He received his Ph.D. in physics and astronomy from the University of Iowa, where he studied under James Van Allen. Hansen is best known for his testimony to congressional committees in the 1980s that helped to raise awareness of the global warming issue.

David G. Hawkins is director of the Climate Center at the Natural Resources Defense Council (NRDC), where he has worked on air, energy and climate issues for 35 years. Hawkins serves on the boards of many bodies that advise government on environmental and energy subjects.

Howard Herzog, a principal research engineer at the Massachusetts Institute of Technology Energy Laboratory, is the primary author of a 1997 U.S. Department of Energy White Paper on carbon sequestration.

John B. Heywood is Sun Jae Professor of Mechanical Engineering and director of the Sloan Automotive Lab at the Massachusetts Institute of Technology. He was educated at the University of Cambridge and at MIT, where he joined the faculty in 1968. He is author of the widely used textbook *Internal Combustion Engine Fundamentals* (McGraw-Hill, 1988) and is a member of the National Academy of Engineering and the American Academy of Arts and Sciences.

William Hoagland received an MS degree in chemical engineering from the Massachusetts Institute of Technology. After working for Syntex, Inc., and the Procter & Gamble Company, he joined the National Renewable Energy Laboratory (formerly the Solar Energy Research Institute) in Golden, Colorado, where he managed programs in solar materials, alcohol fuels, biofuels and hydrogen. Hoagland is currently president of W. Hoagland & Associates, Inc., in Boulder.

Nick Holonyak, Jr., professor of electrical and computer engineering and physics at the University of Illinois, is credited as being the inventor of the first practical LED: the red gallium arsenide phosphide LED. He was also two-time Nobelist John Bardeen's first Ph.D. student.

Eberhard K. Jochem is professor of economics and energy economics at the Swiss Federal Institute of Technology (ETH) in Zurich and director of the Center for Energy Policy and Economics there. Educated as a chemical engineer and economist at the technical universities of Aachen and Munich, he was a postdoctoral fellow at the Harvard School of Public Health in 1971 and 1972 before beginning his research in energy and material efficiency at the Fraunhofer Institute for Systems and Innovation Research. He is a member of the editorial board of several scientific journals and of the Encyclopedia of Energy and a member of the Swiss Academy of Engineering Sciences.

Olav Kaarstad, principal research adviser in the area of energy and environment at the Norwegian oil and gas company Statoil, is currently involved in the ongoing carbon dioxide–injection project at the Sleipner field in the North Sea.

Daniel M. Kammen is Class of 1935 Distinguished Professor of Energy at the University of California, Berkeley, where he holds appointments in the Energy and Resources Group, the Goldman School of Public Policy and the department of nuclear engineering. He is founding director of the Renewable and Appropriate Energy Laboratory and codirector of the Berkeley Institute of the Environment.

David W. Keith is an assistant professor in the department of engineering and public policy at Carnegie Mellon University. He often collaborates on environmental policy research with Edward A. Parson.

Frederick A. Kish, Jr., is a leader in his work on light-emitting diodes. He is R&D and manufacturing department manager at Agilent Technologies, where he was one of the primary instigators of a new family of high-brightness red-orange-yellow LEDs, which were the first LEDs to exceed the efficiency of unfiltered incandescent bulbs and are now the dominant technology for traffic signals and exterior lights on automobiles. Kish and coauthor M. George Craford were graduate students in Nick Holonyak's laboratory.

John F. Kotek has played a leading role in the U.S. nuclear energy program. He is manager of the special projects section at Argonne National Laboratory–West in Idaho and a member of the team that directs the DOE's Generation IV effort. Before joining Argonne in 1999, he was associate director for technology in the DOE's Office of Nuclear Energy, Science and Technology.

Jean H. Laherrère has worked in the oil industry for more than 40 years. His early work on seismic refraction surveys contributed to the discovery of Africa's largest oil field. At Total, a French oil company, he supervised exploration techniques worldwide. He is currently associated with Petroconsultants in Geneva.

James A. Lake has played a leading role in the U.S. nuclear energy program. He is associate laboratory director for nuclear and energy systems at the U.S. Department of Energy's Idaho National Engineering and Environmental Laboratory (INEEL), where he heads up research and development programs on nuclear energy and safety as well as renewable and fossil energy. In 2001 he served as president of the American Nuclear Society.

Daniel A. Lashof is science director and deputy director of the NRDC's Climate Center, at which he has focused on national energy policy, climate science and solutions to global warming since

1989. Before arriving at the NRDC, Lashof developed policy options for stabilizing global climate at the U.S. Environmental Protection Agency.

Alan C. Lloyd is chairman of California's Air Resources Board, part of the state's Environmental Protection Agency. When his article was commissioned, he was executive director of the Energy and Environmental Engineering Center at the Desert Research Institute in Nevada. Before that, he was chief scientist at the South Coast Air Quality Management District in California. He wishes to thank the U.S. Fuel Cell Council and Fuel Cells 2000 for their help in the preparation of his article.

Amory B. Lovins is cofounder and chief executive of Rocky Mountain Institute, an entrepreneurial nonprofit organization based in Snowmass, Colorado, and chairman of Fiberforge, an engineering firm in Glenwood Springs, Colorado. A physicist, Lovins has consulted for industry and governments worldwide for more than 30 years, chiefly on energy and its links with the environment, development and security. He has published 29 books and hundreds of papers on these subjects and has received a MacArthur Fellowship and many other awards for his work.

Gerald E. Marsh is a fellow of the American Physical Society and worked as a consultant to the U.S. Department of Defense on strategic nuclear technology and policy in the Reagan, Bush and Clinton administrations. He is coauthor of *The Phantom Defense: America's Pursuit of the Star Wars Illusion* (Praeger Press).

J. Byron McCormick is executive director of GM's Fuel Cell Activities. He has been involved in fuel-cell research throughout his career, initiating and then heading the Fuel Cells for Transportation program at Los Alamos National Laboratories before joining GM in 1986.

Alex P. Meshik began his study of physics at St. Petersburg State University in Russia. He obtained his Ph.D. at the Vernadsky Institute of the Russian Academy of Sciences in 1988. His doctoral thesis was devoted to the geochemistry, geochronology and nuclear chemistry of the noble gases xenon and krypton. In 1996 Meshik joined the Laboratory for Space Sciences at Washington University in St. Louis, where he is currently studying, among other things, noble gases from the solar wind that were collected and returned to the earth by the Genesis spacecraft.

Ernest J. Moniz is a professor of physics at the Massachusetts Institute of Technology and cochaired the 2003 interdisciplinary MIT study entitled *The Future of Nuclear Power*. He was associate director for science in the Office of Science and Technology Policy (1995–1997) and undersecretary of energy (1997–2001). He is currently cochairing an MIT study on the future of coal.

George Musser is a staff writer at *Scientific American*.

Eric Niiler is a journalist based in San Diego.

Joan Ogden is professor of environmental science and policy at the University of California, Davis, and codirector of the Hydrogen Pathways Program at the campus's Institute of Transportation Studies. Her primary research interest is technical and economic assessment of new energy technologies, especially in the areas of alternative fuels, fuel cells, renewable energy and energy conservation. She received a Ph.D. in theoretical physics from the University of Maryland in 1977.

Thomas J. Overbye, who holds the Fox Family Professorship in Electrical and Computer Engineering at the University of Illinois at Urbana-Champaign, contributed to the official investigation of the 2003 North American blackout.

Stephen W. Pacala leads the Carbon Mitigation Initiative at Princeton University with coauthor Robert H. Socolow. The initiative is funded by BP and Ford. He investigates the interaction of the biosphere, atmosphere and hydrosphere on global scales, with an emphasis on the carbon cycle. He is director of the Princeton Environmental Institute.

Edward A. Parson is an associate professor at the John F. Kennedy School of Government at Harvard University. He often collaborates on environmental policy research with David W. Keith.

Paul Raeburn covers science and energy for *Business Week* and is the author of *Mars: Uncovering the Secrets of the Red Planet* (National Geographic, 1998).

Jeffrey Sachs is director of the Earth Institute at Columbia University and of the UN Millennium Project.

Gunjan Sinha is based in Berlin.

Robert H. Socolow is professor of mechanical and aerospace engineering at Princeton University. He teaches in both the School of Engineering and Applied Science and the Woodrow Wilson School of Public and International Affairs. A physicist by training, Socolow is currently coprincipal investigator (with ecologist Stephen Pacala) of the university's Carbon Mitigation Initiative, supported by BP and Ford, which focuses on global carbon management, the hydrogen economy and fossil-carbon sequestration. In 2003 he was awarded the Leo Szilard Lectureship Award by the American Physical Society.

George S. Stanford, whose research focused on experimental nuclear physics, reactor physics and fast-reactor safety, is coauthor of *Nuclear Shadowboxing: Contemporary Threats from Cold War Weaponry* (Fidlar Doubleday).

Chauncey Starr, a 1990 recipient of the U.S. National Medal of Technology, did early research on cryogenics, managed the atomic energy division of Rockwell International, cofounded the American Nuclear Society, and was president of EPRI for more than a decade.

Gary Stix is an editor and writer for *Scientific American*.

Julie Wakefield is a science and technology writer based in Washington, D.C.

Matthew L. Wald is a reporter at the *New York Times*, where he has been covering energy since 1979. He has written about oil refining; coal mining; electricity production from coal, natural gas, uranium, wind and solar energy; electric and hybrid automobiles; and air pollution from energy use. His current assignment is in Washington, D.C., where he also writes about transportation safety and other technical topics.

Robert H. Williams is a senior research scientist at Princeton University, which he joined in 1975. At the university's Princeton Environmental Institute, he heads the Energy Systems/Policy Analysis Group and the Carbon Capture Group under the institute's Carbon Mitigation Initiative (which is supported by BP and Ford).

Wendy Williams, based in Mashpee, Massachusetts, is studying technologies that reduce carbon emissions through a grant from the Fund for Investigative Journalism.

---

*This information was compiled at the time the articles were originally published in *Scientific American*; some biographies may not be completely up-to-date.

INDEX

World Petroleum Council, 192
Worldwatch Institute, 193
World Wildlife Fund, 70

X

X-Drive, 156
xenon, 93–96

Y

Yucca Mountain, 100, 111, 112

Z

ZTek Corporation, 119